The Transit of Venus

# The Transit of Venus

HOW A RARE ASTRONOMICAL ALIGNMENT
CHANGED THE WORLD

THE COLLECTED LECTURES OF THE
ROYAL SOCIETY OF NEW ZEALAND
TRANSIT OF VENUS SERIES
BROADCAST ON RADIO NEW ZEALAND

AWA SCIENCE

# Introduction

*—Marilyn Head*

IN 2004 THE ROYAL SOCIETY of New Zealand mounted an unusual expedition, sending three teams of students to England to observe an astronomical phenomenon that no living person had seen: the transit of Venus. For the society, it was part of an audacious bid to capture the minds and hearts of New Zealanders; up until then, few of us were aware that it had been a transit of Venus two centuries earlier that had led to the European discovery and settlement of Aotearoa, and fewer still appreciated the rich cultural and scientific history behind the quest to observe it.

The teams, which had been selected on the basis of winning a short video competition, were issued with 'secret instructions' – paralleling the instructions George III had given to Captain James Cook, ordering him seek out and claim the putative Great Southern Continent. The students' mission was more modest: to seek out and evaluate the site for a new New Zealand war memorial in London. Evidently they approved, for the memorial has since been erected.

At home, New Zealanders were treated to these public lectures and broadcasts, and to daily video updates of the students' 'explorations' of the Royal Society, the British and Science Museums, Kew Gardens, Oxford, Cambridge, the Jodrell Bank Observatory, and Cook's home town of Whitby, culminating on 8 June 2004 with a live video broadcast of the transit itself, beamed from Whitby to a rapt audience at the Museum of New Zealand Te Papa Tongarewa.

Who could have imagined that the slow progress of a small black dot across the face of the sun would capture the imagination of a whole nation? And not just any nation, but New Zealand, a land of 'doers' not watchers, and one not much given to retrospective musings – triumphs of All Black rugby teams excepted. That it did was the remarkable achievement of the Royal Society's communications manager Glenda Lewis, whose brainchild the project was. Through the generous support of Freemasons New Zealand, Radio New Zealand, the Royal Navy of New Zealand and the British High Commission, thousands of people were awakened to the pivotal role that the transits – and science – had played in New Zealand's history. Even better, they were awakened to the fascinating advances of modern science and the passionate enthusiasm of scientists.

And they've stayed awake. The huge public interest generated by the project has been reflected in vastly increased media coverage of science, and has indirectly led to such arts-science interactions as 'Are Angels Okay?' in which leading writers and scientists engaged in private dialogue and wrote about each other's disciplines. Another legacy is the superb interactive website, *www.transitofvenus.auckland.ac.nz*. Hosted by the University of Auckland, the site features videos, interviews, and comprehensive cultural and scientific information about the transits, Cook's voyages, Polynesian exploration and New Zealand astronomy.

The short history of observed transits of Venus, where the shadow of the planet is silhouetted against the sun's brilliant disc, abounds with tales of passion and adventure quite at odds with the modern-day perception of science as a purely

The transit of Venus of 8 June 2004, photographed using the same method as that used by 20-year-old Jeremiah Horrocks to observe the transit 365 years earlier, in 1639: a small telescope was set up to project an image of the sun on to a screen. DUNCAN STEEL

intellectual pursuit. Suspense and intrigue, tragedy and triumph, swashbuckling capt'ns and luckless astronomers, war, pestilence, robbery and murder, and above all a glittering prize for which the greatest minds in the most advanced nations competed for centuries: no wonder we were captivated.

First light, to use an astronomical expression, came 400 years ago with Johannes Kepler's laws of planetary motion, which destroyed the traditional notion of perfectly circular orbits and revealed that planetary orbits are ellipses, with the sun at one focus; that their orbital speed varies according to their distance from the sun; and, most importantly, that the square of the orbital period is directly proportional to the cube of the average distance from the sun.

It was this third law that established the relationship between the distance of the planet from the sun and the time it took the planet to complete an orbit, and gave astronomers a unit with which to gauge the size of the solar system for the first time. The distance between the sun and the Earth is one Astronomical Unit (AU). Using the same unit of measurement, Venus is 0.73 AU from the sun and Jupiter 5 AU – the relative distances of the planets from the sun can be calculated from their orbital periods. However, the actual length of an AU was still a mystery: it was like having a map without a scale.

Nevertheless, such understanding made it possible for Kepler to calculate and predict planetary positions with an accuracy ten times better than had been achieved before, and to finally lay to rest the Ptolemaic model of an Earth-centred universe.

It also allowed him, in 1627, to make the first prediction of a planetary transit. Astronomers use a number of terms to describe what happens when one celestial body passes between two others. When the moon passes between the Earth and the sun, this is called an eclipse; when the moon passes in front of a star or planet, it is an occultation; and when a small body passes in front of a large one, for instance when one of Jupiter's moons is seen travelling across the face of the planet, it is a transit.

From Earth, Venus and Mercury are the only planets that transit the sun, as they are the only ones whose orbits are 'inferior' to, or inside, Earth's. Kepler predicted that in 1631 Venus would transit the sun, and that a second transit would occur the following century, in 1761.

In keeping with his abysmal luck, this brilliant, almost blind German mathematician, who had lost his family to smallpox,

Johannes Kepler, born in Weil der Staadt in the Holy Roman Empire in 1571, was a passionate defender of the theory of Copernicus that the planets orbited the sun rather than the Earth, and became the first person to accurately explain their motion.

who had been besieged and driven from home after home by the religious and political conflicts that plagued the Holy Roman Empire and eventually led to the Thirty Years War of 1618–48, and who had never consistently received the royal patronage he had been promised, did not get the chance to see his predicted transit. He died the year before, fleeing yet another war-torn city. In the event, although it was looked for, the sun had set in Europe before the transit began, and the event was not seen from anywhere else.

Enter another brilliant mathematician, but one with keen eyesight as well as a talent for astronomical observations. As passionate a fan of Kepler as any modern teenager is of a favourite rock band, Jeremiah Horrocks, a young, impoverished tutor, immersed himself in Keplerian computations before coming to the surprising conclusion that his hero had missed predicting a second transit of Venus on 24 November 1639. In fact, because of the way the orbits of Earth, Venus and the sun coincide, transits occur in pairs separated by eight years every 120 years or so.

With little time to advertise the fact, Horrocks enjoined another amateur astronomer, William Crabtree, to observe the transit in Manchester, while he observed it from Carr House, near Much Hoole in Lancashire. Both used their small home-made telescopes to project an image of the sun on to a paper screen.

'Rapt in contemplation, I stood for some time motionless, scarcely trusting my own senses, through excess of joy; for we astronomers have, as it were, a womanish disposition, and are overjoyed with trifles ... as scarcely make an impression on others.' Such were Crabtree's thoughts as he successfully observed the transit with his family.

Horrocks, the only person in the world who had calculated when the event would occur, was less fortunate, his observations being hampered by clouds. But these eventually parted to reveal a 'most agreeable spectacle' and allow him to partially time the transit. Both men were amazed at the disparity in size between the tiny dot that was Venus and the vastness of the sun's orb.

Horrocks's subsequent calculations refined both the orbit of Venus, and the estimate of the mean distance between the Earth and the sun. Although his estimate of 59 million miles (95 million kilometres) for the length of the AU was about two-thirds of what we now know to be the value, it was the most accurate figure to date. It is not merely facetious to add,

Jeremiah Horrocks, an impoverished 20-year-old mathematics tutor, correctly predicted that Venus would cross the face of the sun on 24 November 1639. As pictured in this 1903 painting, he observed the event using a home-made telescope to project the image on to a paper screen. It is dangerous to look at the sun directly. This romantic Victorian painting of Horrocks, entitled *The Founder of English Astronomy*, is by Eyre Crowe.

'Alas, poor Horrocks!', because this mathematical genius, whose work on gravity predated Newton's by a quarter of a century, died two years later at just 22.

The next player in the transit saga was Edmond Halley, and anyone with a less 'womanish disposition' would be difficult to imagine. Forget the bewigged and portly figure most of the portraits show and imagine the maritime adventurer who commanded a pink, the *Paramore*, and swore like the sea captain he was. Prodigiously talented and energetic, in 1676, aged barely 19, Halley led an expedition to St Helena, an island off the coast of Africa (where Napoleon Bonaparte would later breathe his last), in order to catalogue far southern stars.

While there he observed a transit of Mercury, which inspired him to suggest a method by which a more precise measurement of the Astronomical Unit could be obtained. (If that doesn't sound too exciting, remember that at the time the sun had barely been given its rightful place in the centre of the solar system, and it would be another 200 years before the Milky Way was recognised as just one galaxy among trillions.)

Halley realised that if transits of Venus were to be simultaneously timed from widely separated geographic locations, it would be possible to measure the parallactic shift of the planet against the sun. Parallax is the shift you see when you hold your thumb up in front of a distant object and look at it with first one eye and then the other; your thumb appears to move against the background. A simple geometric calculation could then be used to determine the actual distance to Venus, and thus the distance to the sun – the Astronomical Unit. This was the astronomers' Holy Grail, the absolute number that would allow them to at last find out just how big the solar system was.

In spite of knowing he would not be around to see them (just as he would never see the return of the comet that bears his name), Halley urged the scientific community to mount global expeditions for the next transits in 1761 and 1769. Thus began the 'great chase after Venus's shadow', as American astronomy historian William Sheehan has called it. Coinciding with a period of intense Anglo–French rivalry, it was, in Sheehan's words, 'the eighteenth-century equivalent of our own Race to the Moon'.

In 1761, England, France and Austria joined the chase, and there were also observers throughout Europe. However, many were hampered by war, weather and the 'black drop effect', a curious observational distortion that made precise timing especially difficult. To understand both the cause and effect, try bringing your thumb and forefinger together and determining the precise moment of contact. You will notice that the point of contact appears smeared out.

Although successful observations *were* made from sites as far afield as Siberia and South Africa (the latter by Charles Mason and Jeremiah Dixon, who later surveyed America's Mason-Dixon line), the timings lacked precision because of the black drop effect, and because many observers were unable to determine their positions accurately.

All in all, the 1761 observations were a keen disappointment, so the pressure was on for the last opportunity of the century – the transit of 1769. The Royal Society took the lead, planning expeditions to remote places such as Hudson Bay in the Canadian north and newly discovered Tahiti. Not to be outdone, especially by a nation with whom they were at war, the French sent teams to Siberia and South America. The scientific quest became a matter of national pride, but it is a moot point whether it motivated or simply coincided with the extraordinary wave of exploration and colonisation of the eighteenth century. However, the appearance of a small black dot gliding across the face of the sun was at least partly responsible for colonial settlement in such remote and disparate places as the Pacific Islands, Canada, Siberia, Africa, including Madagascar, and India.

Mindful of the difficulties of the 1761 expedition, which had led to less precise measurements than had been hoped for, the Royal Society wanted to ensure not only that its expeditionary forces would reach the right destination – always a tricky business with the inexact state of surveying and navigation – but also that they would be competent enough to pinpoint their geographic location, and to time the transit accurately.

And who better to meet those challenges than James Cook? The captain's mastery of the astronomical instruments used for surveying and charting unexplored territories was such that he was paid an additional 100 guineas to act as one of the two professional astronomers charged with observing the transit of Venus from Tahiti. Cook's barque *Endeavour*, as familiar to New Zealanders as Columbus's *Santa Maria* and the Pilgrims' *Mayflower* are to Americans, set sail in 1768, arriving in Tahiti a few months before the transit was to take place on 3 June – time enough to establish temporary observatories and test the equipment, although the latter proved a temptation to a Tahitian or two.

Fortunately the weather cooperated and the observations were successful, but once again the black drop effect introduced an unusually large margin of error and Cook was greatly disappointed in the disparity of his own and astronomer Charles Green's timings.

He need not have been. By 1771 Thomas Hornsby, professor of astronomy at Oxford, had averaged the plethora of observations from all over the world, eliminated suspect data and deduced an Earth-to-sun distance of 93,726,900 miles (151 million kilometres), remarkably close to today's accepted value.

Cook sailed on to Aotearoa, taking the Tahitian navigator and chief Tupaea with him as guide and translator, and took 'formal possessions of the place in the name of His Majesty'. He observed a transit of Mercury in the Coromandel, before spending six months mapping the entire New Zealand coastline.

THE FOLLOWING CENTURY saw an alternative method of determining the distance to the sun, using the aberration of starlight. This is the very small change in the apparent position of stars caused by the Earth's movement around the sun. The movement, which had baffled astronomers, was explained by the British Astronomer Royal, James Bradley, in 1829. Bradley realised that the apparent position of the stars was due to the combined effect of the speed of light and the motion of the Earth. Indeed, the aberration of starlight was the first direct evidence that the Earth *did* move around the sun.

By measuring the aberration of starlight it was possible to calculate the Earth's orbital speed. Knowing the speed and the orbital period meant the circumference of the orbit could be calculated, and thus the radius – the elusive AU.

Problem solved? Not quite. Today we know that the speed of light is a universal constant, but Bradley did not. In the nineteenth century there was still room for doubt about the accuracy of the measurements derived using this method.

Determining an accurate value for the AU became even more pressing a decade later, when the first stellar parallax was measured. A reliable distance to a nearby star had been made, but astronomers were still in the dark about the size of the solar neighbourhood. Armed with better telescopes, 'smoked glass' filters that reduced the transmission of harmful solar radiation (they were made by depositing soot on the glass and had to be handled very carefully), heliometers for measuring the angular diameter of the sun, essential for accurate timings, and photographic capability, the astronomical community was spurred to mount numerous expeditions all over the world for the 1874 and 1882 transits.

But what a different world it now was. Far from observing the transits in isolation and among 'alien' cultures, most astronomers were greeted enthusiastically by thriving communities where many people understood the purpose of the observations and evinced a passion for scientific pursuits. Education and technology allowed non-scientists to share in the thrill of a scientific adventure that took them beyond the confines of the planet, and the transits generated huge public interest.

Novels such as Thomas Hardy's romantic *Two on a Tower* were written, poems were penned, music was composed, and numerous scientific lectures given. The transits were observed by amateurs and professionals alike, brokers in New York queuing up to pay 10 cents for a 15-second peek. In New Zealand, British and American teams were sent to Queenstown, Burnham near Christchurch, the Chatham Islands, and what became known as Observatory Hill in Auckland and is now the site of the War Memorial Museum. A German team even spent several months in the inhospitable subantarctic Auckland Islands.

As usual, the weather took its toll and not all observations were successful, but many thousands of photographs and more precise timings gave rise to years of analysis. Again, a reasonably satisfactory result was obtained using a weighted average, not only of the transit data but also of that obtained by other methods.

There were no transits in the twentieth century, but the century nevertheless saw unprecedented advances in both theoretical and experimental astrophysics. Technology followed. Soon astronomers found that other solar system objects could be used to derive a solar parallax, and thus a precise mean

Earth-to-sun distance. In 1931, astronomers used the very close approach of the asteroid Eros to photograph and make micrometric measurements of its position in relation to nearby stars. The results implied an average AU of 149,675,000 kilometres, with a margin of error of only 17,000 kilometres.

Later, radio signals bounced first off the moon and later off Venus were timed, yielding measurements within a few hundred kilometres. Telemetry from space probes such as the *Viking Lander*, on its mission to Mars, reduced that to a few metres. Horrocks and Halley's technique for measuring the scale of the solar system had been well and truly superseded.

However, in a serendipitous twist of fate, the transits had not reached a scientific dead end. By 2004, the time of the first transit of Venus in the twenty-first century, astronomers had found a new reason for observing them. A planet that transits the face of a star causes the star to dim by a very tiny fraction and this, in conjunction with the radial velocity technique, is a way in which extra-solar planets have been detected. Indeed, New Zealand astronomers Denis Sullivan and his son Tiri Sullivan are among the few to have detected an extra-solar planet this way: in 1999, using a photometer they had built themselves, they recorded the tiny drop in luminosity of the star HD209458 from an observatory atop Mauna Kea in Hawai'i. The method allows the orbital size to be determined from the period as well as the mass and size of the planet (from Kepler's third law and the depth of the dip in the light curve, respectively). Most importantly, it offers a sensitive probe of planetary atmospheres.

Several space missions, including the French *Corot*, which was launched on 27 December 2006, and the United States'

*Kepler*, due to be launched in November 2008, have been designed to survey hundreds of thousands of stars for planets using this 'planetary transit' technique.

ALTHOUGH WE CAN NOW determine the mean Earth-to-sun distance to much better than one part in 100 million (about 150 million kilometres or 149,587,953 to be more precise), the rich history and science of the transits still excite keen interest. I was privileged to see the 2004 transit of Venus from Carr House in Much Hoole, from the exact spot where Jeremiah Horrocks first observed the 1639 transit. There may have been a jazz band in the garden and the paraphernalia of television crews and modern astronomical equipment, but my feelings were the same as his had been: it was indeed a 'most agreeable spectacle'. To see the tiny black dot that was Venus, a planet as big as our own, kiss the limb of the sun at exactly the predicted time and glide across the disc, before disappearing, was both thrilling and humbling. Thrilling because of the intellectual rigour and imagination that has allowed humans to penetrate some of the mysteries of a vast and complex universe; humbling because it was a graphic demonstration that our whole planet, and all the species and people who live or have ever lived on it, amount to no more than a tiny dot, orbiting an ordinary star, in one of hundreds of billions of galaxies.

As astronomer Carl Sagan observed: 'Our planet is a lonely speck in the great enveloping cosmic dark ... To me it underscores our responsibility to deal more kindly with one another and to preserve and cherish the pale blue dot, the only home we've ever known.'

In 2012, New Zealanders will have the once-in-a-lifetime opportunity to see the last transit of Venus this century, and to contemplate the role that the transits, and particularly the transit of 1769, played in the development of our and other nations. In the meantime, the entertaining and informative essays in this book provide a tantalising introduction to the geological, astronomical, historical, anthropological and physical importance of these rare and captivating astronomical alignments.

# Search for the Lost Continent

*—Hamish Campbell*

YOU MAY WONDER WHY a geologist is talking about the transit of Venus. I shall explain that simply. Captain Cook had secret orders that, having observed the transit of Venus, he should move south and west in search of land, in particular a continent. I am going to be describing one of the places he found – New Zealand: how it got here, how it split away from Gondwanaland, and the implications of this for the development of its unique fauna and flora.

Let me tell you straight away that what Cook discovered was indeed a continent – or rather its submerged tip. He didn't know this at the time, for all sorts of reasons. If you read the journals of Cook and his companions Banks and Solander, you won't find any mention of 'geology'. The word did not exist: geology was not yet a scientific discipline. Indeed the year Cook arrived in the Pacific, 1769, saw the birth of William Smith, the English surveyor who would become one of the fathers of geology. But this lay in the future: at the time of Cook's voyages the state of geological wisdom was low. The general understanding was that the world had been created, according to one of the vice-chancellors of Cambridge University, at 9 a.m. on September 17, 4004 BC.

Less than 150 years later, New Zealander Ernest Rutherford would revolutionise our understanding of the age of the Earth when he wrote a paper in which he suggested that it should be possible to determine the age of minerals, and hence of rocks. These days we can date virtually anything that happens to have a radioactive element in it. A few weeks ago I was visited by a colleague from Münster, Germany who has been working on some of the oldest objects we know of in the solar system. I asked him, 'Am I still correct in stating the age of the Earth at 4.6 billion years?' He very earnestly replied that that was completely incorrect. The Earth was, in fact, 4.53 billion years old. He then went on to tell me that the age of the moon is 4.45 billion years.

Now, we've all heard that New Zealand split away from Gondwanaland. But what is Gondwanaland, and why did the split happen? First, let me make it clear that Gondwanaland was a continent – that is, by definition, a land area greater than about 100,000 square kilometres. It gets its name from the label given to a common sequence of rock that occurs in the land masses of Australia, Africa, India, South America and Antarctica. The term was coined by a British geologist, H.B. Medlicott, who worked for the Geological Survey of India in the 1870s. Gondwana was a kingdom in India, located south-west of Calcutta, which had been overcome by the Mogul empire in the 1500s. It was in this region that the Gondwana sequence was first described.

Before Gondwanaland was formed, there were other configurations of continental land masses. We know that a continent is a major element of the Earth's crust, but what is the Earth's crust? Well, think of Planet Earth as being like your body – an entity with an internal apparatus and an outer crust. The crust is really like a skin or tissue, and is very, very thin compared to the Earth's radius. While Earth's radius approaches 6500 kilometres, most of Earth's crust is only seven kilometres thick.

Most of this crust lies under the oceans, which occupy more than 60 percent of the surface of the planet. And in the late

1960s scientists determined, to their amazement, that nowhere on Earth is the sea floor beneath the oceans older than about 180 million years, and that it is all comprised of one rock type, basalt. This was an astounding discovery: if you scraped or dug beneath the sediment on the sea floor, you came to basalt and none of it was older than 180 million years. By this time a great deal of dating had been done on the continents, and the oldest continental rocks had been found to be about 3.8 billion years old. How could you have an ocean floor that was so much younger than the continents?

The answer turned out to be that the ocean floor is continually being replaced. It arises, it forms, it spreads and then it sinks back down. This is referred to as sea-floor spreading, and its operation is described by a theory called plate tectonics. But how could this possibly happen? The world is surely made up of rock, and you can't move rock easily. As you can imagine, an awful lot of people thought this idea was just ridiculous.

I now want you to consider what Planet Earth looks like. Think of a map of the world with blue sea and land. I am going to use an analogy: imagine that the blue sea is milk, and the continents are cream. This will help you understand how the crust works.

In the late 1960s, geologists established that there were two types of crust: oceanic, composed primarily of basalt (the milk); and continental, composed primarily of granite (the cream). Basalt is denser than granite and is dominated by iron- and magnesium-bearing minerals, whereas granite is dominated by silicon- and aluminium-bearing minerals.

These are two very distinct types of rock, with different densities because of their different compositions. And the amazing thing is that they are both igneous – that is, they start from a molten state. In a sense, then, we can describe the Earth's crust as being made up of two frozen liquids. These liquids come from the Earth's mantle, which forms the great bulk of the planet between the core and the crust. From the moment the Earth formed billions of years ago, its mantle has been hot and mobile – 'at work', so to speak, evolving and generating these two common fluids, basalt and granite, that rise to the surface and contribute to the crust.

There has probably never been as much granite as there is now forming continental crust on the surface of the Earth. Geologists can trace the evolution of continental crust back about three billion years, and as the mantle has evolved, granite has been slowly accumulating. Prior to three billion years ago, there was no crust and no continents.

Let's now think about oceanic crust (basalt). Today, new oceanic crust rises as liquid rock (referred to as magma at depth, and lava at the surface) along the mid-oceanic ridges that straddle the entire globe, as a connected line through all the oceans for about 65,000 kilometres. Mid-oceanic ridges are almost exclusively submarine. There are very few places on Earth where an active mid-ocean ridge can be seen on land; one, and the best example, is Iceland.

The whole system works like a giant conveyor belt. There are basically just two fluids, with one – the milk, or oceanic crust – driving the other. The cream, or continental crust, is literally sitting on top, changing shape and being driven around the place by the motion of the milk, which in turn is being driven by convective motion within the Earth's mantle. It is a very simple process.

Now, to return to the question of why Gondwanaland split up and produced New Zealand, again the answer is within the mantle. To understand this, think again of your body. The nature of your skin is determined largely by the processes happening within your body, and the same is true of the Earth's crust: almost everything is being controlled by processes within the mantle. And what broke away at a seminal point about 85 million years ago was Zealandia, a continent about one-third the size of present-day Australia. My colleagues in GNS Science would say it broke away precisely 83 million years ago: that is the age of the oldest oceanic crust beneath the Tasman Sea.

I want you to visualise this great rift occurring, the Tasman sea floor forming, and Zealandia, the continent, being drawn away. As it does so, it both sinks *and* drifts away from the thermal hotspot from which fresh oceanic crust is emanating and forming the floor of the Tasman Sea. And as this New Zealand continent, Zealandia, moves away, it carries with it a cargo of plants and animals. We can be positive about this, not least because of the wonderful discovery in 1980 by amateur palaeontologist Joan Wiffen of the first dinosaur fossils in New Zealand. Wiffen's crowning achievement was not recognising one dinosaur fossil, but establishing that there had been a whole community of dinosaurs. Today we know of six species, all of which lived about 75 million years ago.

Zealandia, with its dinosaurs and plants, drifted away from the Australian and Antarctic margins of Gondwanaland for about 20 million years. For about 65 million years since then, this continent (and hence also New Zealand, which is a part of it) has been essentially fixed in its present location with respect to Australia.

Why did the sinking occur? One can understand a continental land mass sinking as it drifts away from the hot, thermally inflated process that has caused it to split and grow colder. But this process went on for about 60 million years, reaching a low point about 25 million years ago. The geological record of New Zealand tells us this: all over the country there are remnants of a blanket of limestone. Limestone is a sedimentary rock that accumulates on the sea floor and is biogenic – that is, almost totally composed of the skeletal remains of marine life. From this evidence we can be reasonably certain that Zealandia – and the New Zealand land mass as we know it today – were once almost totally submerged, with the possible exception of a few small islands.

So we have this amazing history: Gondwanaland breaking up, and Zealandia moving away, sinking, and reaching a low point about 25 million years ago. And then what? Then the modern plate boundary kicked into action.

The Earth's crust consists of about 15 large plates, each of which are made up of both oceanic and continental crust. And while oceanic crust is about seven kilometres thick globally, continental crust is much thicker. And the thicker the continental crust, the higher it will sit with respect to the Earth's surface. Hence, the highest places on Earth – the Andean plateau, Altiplano, in South America and the Tibetan plateau in Asia – are supported by 70 to 80 kilometres of continental crust. In New Zealand, however, the crust is anomalously thin, only 25 kilometres on average. And that is why our continent, Zealandia, is largely submerged.

I want you to imagine a bathymetric map of the New Zealand region – one that shows water depth. Everything shallower than

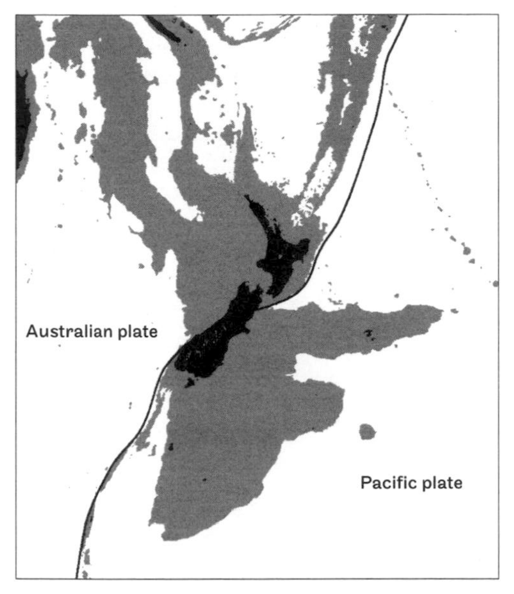


The land of New Zealand is merely the emergent part – just seven percent – of the sunken continent of Zealandia. Uplifted by a collision along the boundary (represented by the black line) between the Australian and Pacific plates, it is in a constant state of change. The extent of Zealandia is roughly determined by water depth: below 2500 metres is dense, less buoyant oceanic crust; above 2500 metres is less dense, more buoyant continental crust.

about two and a half kilometres will indicate continental crust and define the land area of Zealandia. Twenty-five million years ago most of New Zealand was submerged like this. Then the plate boundary became vigorous and – bingo! – the whole lot was pushed back up. This collision has been going on ever since, and thank goodness, because otherwise New Zealand would be sinking below the waves. The ambient level at which Zealandia wishes to be is one to two kilometres under the sea.

The nature of the collision varies dramatically up and down the country. The collision in the North Island is between oceanic crust on the Pacific plate and continental crust on the Australian plate (cream v. milk). The whole of the North Island is on the Australian plate, as is Nelson and the top end of the South Island. The boundary between the Pacific plate to the east and the Australian plate to the west runs from north-east to south-west, entirely offshore down the eastern side of the North Island. It then bends in a more east–west direction in the Cook Strait region, before cutting across the South Island.

If you depict the plate boundary as a single line crossing the South Island, you would pick up the Hope Fault just north of Kaikoura, along the southern margin of the seaward Kaikoura Range. This is the most active of several major faults in Marlborough that are sub-parallel to one another, others being the Wairau, Awatere and Clarence. The Hope Fault connects with the Alpine Fault, which runs down much of the West Coast on land along the western margin of the Southern Alps, before running offshore at the entrance to Milford Sound, and down into the Puyseger Trench. So that's the boundary.

In the South Island the nature of the collision is very different. Here it is basically a collision between continental crust and continental crust (cream v. cream), much as you have in the Himalaya or the European Alps. South of Milford Sound the nature of the collision changes yet again, to cream v. milk, but now it is between continental crust on the Pacific plate and oceanic crust on the Australian plate. This is the reverse of the situation in the North Island.

This variation in the nature of the crust involved in the collision along the plate boundary accounts for most, if not all,

New Zealand's geomorphology. It does not matter where you are, the landscape and geology can be explained in terms of this extraordinary collision.

Now, remember we had a cargo – including dinosaurs. And we can be sure there were forests and swamps, because we have coal measures that can only have accumulated and formed on land between 70 million and 45 million years ago. We can see remnants of these coal measures all over the country, and we know of them in the sub-surface of the sunken continent of Zealandia.

So there's no doubt about the land and the fauna and the flora: they were here. However, we have some interesting questions about the fauna and flora. We've always assumed that all our iconic plants and animals must be directly descended from those that came from Gondwanaland with Zealandia. We are now beginning to suspect it may be a lot more complicated than that. It is just conceivable that a great deal of Zealandia's (and hence New Zealand's) fauna and flora was effectively wiped out around that 'low point' about 25 million years ago, and that most, if not all, of the plants and animals with which we are now familiar got here subsequently. A great deal of modern molecular biological work tends to support this contention. Much of New Zealand's fauna and flora is, therefore, almost certainly derived from much younger ancestors than we have previously thought.

This puts a very different perspective on New Zealand's history.

You may be interested to know that the most stable part of New Zealand-Zealandia is the Chatham Islands. The Chathams are almost 1000 kilometres away from the plate boundary

One thousand kilometres away from the boundary of the Australian and Pacific plates, the Chatham Islands are the most stable part of New Zealand-Zealandia. This forest on Rangatira Island in the Chathams is characteristic of what Zealandia looked like before it drifted away from Gondwanaland. CRAIG POTTON

where all the action takes place, so little has happened to them since Zealandia split away from Gondwanaland. If you want to see what Zealandia looked like when it drifted away from Gondwanaland, the Chatham Islands is the place.

The New Zealand land surface is surprisingly youthful: most of it – conceivably *all* of it – is less than 25 million years old. Rocks, minerals and fossils are, in effect, the memory banks of planet Earth. Scientists have worked out ways to interrogate these memory banks, and we are now able to readily deduce much amazing information. We know, for example, that the rocks that make up our land mass are much older – the oldest, in Nelson's Cobb Valley, are 510 million years old – and we are able to trace their ancestry using various techniques. We can determine the source of ancient sediments on the basis of their mineral composition and the age of some of those minerals. In particular, modern geology makes use of the relatively common mineral zircon, which is easily dated using radiometric methods (uranium–lead). Using this technique we have found mineral sand grains that are in excess of three billion years old.

Captain Cook may have looked hard and found no continent here, but if he were alive today he would be delighted to know that, in the last few decades, geologists have recognised that there is indeed a continent. It just happens to be submerged.

***Hamish Campbell,*** *a geologist and research scientist with GNS Science, was educated at Otago, Auckland and Cambridge Universities, where he specialised in palaeontology. Much of his research concerns the older rocks and fossils of New Zealand, those that are between 300 and 150 million years old. Another of his research interests is the geology of the Chatham Islands. However, he is best known as the geologist at the Museum of New Zealand Te Papa Tongarewa, where, as a science communicator, he helps promote geology and provide geological information to the public.*

## The Road to Stonehenge

*—Richard Hall*

WHAT IS THE LINK BETWEEN the star lore of ancient peoples and the 1769 voyage of Captain James Cook? The two are more closely connected than you might at first realise: it was the knowledge of the stars that developed among these ancient peoples that ultimately laid the foundations of navigation and astronomy, which led, in due course, to Cook's mission to observe the transit of Venus.

Many people, myself included, tend to take for granted the vast amount of knowledge that is readily available today through computers and books, but for most of human history there was no such thing as the written word. Much of the information upon which civilisation is based was learned and discovered long before the written word existed. And of this knowledge, none was more important than astronomy.

Let us start by taking a journey back in time, to a place on Earth 20,000 years ago. The sun has set, and as the twilight fades a small band of our ancestors huddle around the campfire. As they listen to the crackle of the burning wood they also hear other noises of the night: the roar of the lion and the howl of the wolf. The night is the time of the predator. In darkness we are robbed of our most important sense, vision, and so the campfire gives our ancestors considerable comfort, not only warmth but also light by which they can see. Most importantly, it keeps the predators at bay.

Looking out from the campfire to the boundaries of its light, these people on occasion see the eyes of these predators, shining like jewels in the dark. But they also look upwards and see the stars. Some believe these lights in the sky are the campfires of other beings, wanderers in the heavens. Because of their familiarity, the lights are comforting companions. The patterns they form, which today we call constellations, are just as well-known to these ancient peoples as the hills and valleys of the lands in which they live.

The knowledge these ancestors of ours gain by observing and watching the stars will ultimately lead to our way of life today. Indeed, it is my contention that without a knowledge of the stars and star lore, civilisation itself would never have come into existence.

HOW OLD IS OUR SPECIES, *Homo sapiens*? The oldest known remains, found in Ethiopia, date back 130,000 years. These people had a primitive technology: they had fire, and used simple hand stones as tools, but they certainly weren't armed with bows and arrows or harpoons, or anything similar. Eventually, about 100,000 years ago, they migrated from Africa into what is now Iraq, and then travelled along the equatorial zone, where the climate didn't change too much. And about 50,000 to 60,000 years ago they reached Australia. These early migrations were easier than they would be today, because during the ice ages sea levels dropped by about 120 metres. Many areas that are now islands separated by water were then land masses, and travellers were able to cross them with ease.

However, despite this diaspora their way of life didn't change; it remained on a very simple technological level for tens of thousands of years. Then, around about 40,000 years ago, something astounding happened, which many archaeologists call the 'second wave'. Suddenly, a new people – when I say 'new', they were still *Homo sapiens* and probably didn't differ biologically from us, or from their predecessors, much at all – appeared who had an advanced technology. And these people rapidly overran the rest of the world.

They first arose in what we now call the Middle East. From there they migrated outward into Africa and Asia, often overrunning areas that had been colonised earlier. But they also moved into areas, such as northern Europe, beyond the equatorial zone. We know that 20,000 years ago they travelled across the Bering Strait into the Americas, and about 5000 years ago they began to explore the great oceans of the world, including the Pacific. Just over a thousand years ago they reached Hawai'i and Aotearoa.

Within 40,000 years, then, these early humans had colonised just about every liveable land mass on the planet.

The question is, what happened 40,000 years ago that led to this dramatic human migration? The archaeological evidence suggests that both the culture and technology of these people had taken a sudden and remarkable leap forward. We know they buried their dead ceremonially, leaving artefacts within the graves, which indicates that they had developed a concept of spirituality and an afterlife. They created works of art – engraved bones, sculptured figures, and cave paintings – which indicates they had moved beyond purely mundane concerns. And we know for certain that they cared for each other quite well. One indication of this is a human thigh bone found by archaeologists. The bone carries a terrible scar that was probably inflicted by a wild boar or something similar when the man was young. The wound would have crippled this individual, but the age of the bone shows he lived to a great age, so somebody had been carrying him around for most of his life. In other words, these early people exhibited what we call 'humanity'.

In addition, there was technological change. Our ancestors fashioned finely crafted weapons. They invented the bow and arrow, and the harpoon, and with them became the most deadly predator on Earth: whenever this new breed of *Homo sapiens* appeared in a territory, the number of large animals tended to diminish rapidly. They also used needle and thread, and made footwear and fine clothes. The combination of these skills allowed them to exploit just about every environment on Earth. And all of this started at one time in just one place.

So what *did* happen to bring it about? We don't know for certain, but one suggestion is that these people had developed the use of complex language. Now you might argue that people had language long before that, and indeed we know that chimpanzees and many other animals have a form of language: they have words they associate with objects. However, our species is the only one capable of putting these sounds together into what we call sentences and paragraphs, whereby it is possible to create visions and ideas, and transfer them to somebody else. This development of language may have been due to a small genetic change within a single group. This, in turn, may mean that every person on this planet has evolved from one small family that lived in the Middle East 40,000 years ago – quite a sobering thought.

To return now to the stars: why were they so important to these people, and indeed to the development of civilisation? The first reason was navigation. If you look at a map of the world, the one significant difference between these people and their predecessors is that they migrated across large, inhospitable terrains. They traversed vast deserts of sand, wastelands of ice, and immense oceans.

Imagine that you are a traveller and before you is a desert of flat, featureless plains and drifting sand dunes. There are no roads and no maps, but somewhere out there, beyond the horizon, is an oasis. Find that oasis and you can survive passage across the desert. Miss it and you will perish. How did our ancestors find their way across such terrains, which had no permanent features? The answer is that they followed the guiding stars. Using the stars, the Polynesians even navigated the greatest ocean on Earth.

As well as being aids to navigation, stars provided a way of telling the time. In the equatorial belt, where these people had been living, there had been very little seasonal change and normally an abundance of food, but in the northern territories to which they migrated they had to deal with marked seasonal changes. There, many animals and birds upon whom these hunter-gatherers depended migrated back and forth with the coming and going of the seasons. When animals migrate they do so *en masse*, and they move a lot faster than people can on foot. To ensure that food supplies did not run out, the hunter-gatherers would have had to move to where the game was going, and they would have needed to move before the animal migrations began.

And there was another imperative. The northern summers, as they still are today, were fertile but very short. When winter arrived, it did so with a tremendous harshness. If they left their move too late, the snows would cut off their retreat. But how did they know exactly when to leave? It was not just a matter of waiting for the trees to change colour: they had to have moved out well before that happened.

We know that early Maori in New Zealand observed the rising of particular stars to foretell the changing of the seasons, the sea currents and the winds, and it is likely that these ancient migrants used a similar system. Indeed, if we look at bone carvings by North American Indians, for example, we find star patterns, or constellations.

When agrarian society began to appear – as nomadic peoples settled down, farmed the land and refined animal husbandry – timekeeping became essential. We used to think that settlement hadn't begun until 10,000 years ago, but recently in the Jordan Valley archaeologists have found remnants of settlements going back 19,400 years. These people, cultivating the land, would have had to know when to sow and when to reap. I have a tiny farmlet in the Wairarapa, and if I plant something at the wrong time and it dies, all it has cost me is money. However, if I had done that 10,000 years ago my family would face starvation.

Knowledge of stars that could be used for navigation and stars that foretold seasonal changes, was, therefore, essential. But, in the absence of written language, how was this information stored? Each tribe or extended family would have had a storyteller, a learned person who had been trained from an early age to remember specific information. When called upon, this person could recall the information in detail: they were, literally, living books. Their information was passed down through story, poetry and song. This makes sense. If I ask a group of people, 'Who knows the story of Goldilocks and the Three Bears?', virtually everyone will put up their hands. The stories and fairytales you learn as a child stay with you for life.

What tends to happen, though, is that with the passage of time we remember the stories and forget the knowledge. We then call these stories 'myths'. Personally, I don't believe there is such a thing as pure mythology. Every one of these great stories from antiquity, whether of gods or giants, great battles or amazing journeys, originally contained practical information. As a Maori kaumatua said to a friend of mine, 'Don't tell me facts, tell me a story: tell me a story and I'll remember the facts.'

Another clue to the link between the stars and the cycle of the seasons is the names given to constellations. A constellation is a pattern of stars perceived and named by a group or race of people. The origins of some of today's common constellations go back 15,000 years. A constellation may be mentioned in, say, Greek literature, but have its origins in an earlier era. Beneath the stars at Stonehenge Aotearoa, people often ask me to point out a particular constellation. It is always a sign of the zodiac and I know why they are asking: because they think it is their birth sign. (As an aside, I have to inform them that, due to the Earth's precession, their real star sign is different from the one shown in their daily newspaper or favourite magazine.)

Anyway, they will ask, 'Where is Pisces?' or 'Where is Aquarius?', and when I show them they're disappointed, because the pattern of stars doesn't look anything like the figure it's supposed to represent. The stars of Pisces don't look like a fish; the stars of Aquarius don't look like a water-carrier.

So how, then, did these and the other constellations get their names? The answer has nothing to do with patterns in the sky. The ancient constellations were named for their symbolism. Aquarius is so named not because it looks like a man with a jar, but because 5000 years ago, when these stars rose just before the sun, the event marked the onset of the rainy season. The gods were seen as pouring water down on the Earth. Similarly, the stars that form the constellation of Cancer, the crab, were so named not because they looked like a crab in the sky, but because when these stars rose in the dawn sky all the crabs in the Nile came out on the land. These constellations were marking seasonal events.

Many of the constellations are denoted by animals, and that shouldn't be surprising either, because their rising was invariably related to the migrations of animals. In the northern hemisphere, for example, we have the 'great bear' in the sky. Bears are dangerous predators so it's a good idea to know when they are around, and you can tell that by looking at the stars. The constellation of the great bear moves around the celestial pole. When the bear stars are down, the bears are hibernating, but when the bear stars come up, bears are abroad.

Timekeeping was also important for trade, and the meeting of tribes. Trade routes were established 40,000 years ago, but how did traders arrange to meet up with each other and exchange goods if they didn't have a clock or a calendar? Well, there is a great clock in the sky: we call it the moon. The regular cycle of the moon, from one new moon to the next, takes 29.5 days, or one month. This is where the word month, or 'moonth' comes from. The 'week' also comes from the cycle of the moon: each week represents a quarter of the moon's monthly cycle. It takes around seven days to move from new moon to first quarter, another seven to move from quarter to half moon, and so on. These are the 'weeks' of the 'moonth'. One group of people could, for example, arrange to meet up with another at the full moon three moonths hence. If the journey were to take about

14 days, they would need to set out at the new moon before the third full moon. Hence, long before there were mechanical clocks and calendars, people used the moon as a timepiece.

Another important aspect to stars is their spiritual symbolism. To our ancestors the stars were untouchable, and appeared to be immortal, so it should come as no surprise to find stars at the very foundation of spiritualism and religion. If you watch a particular star rise up and this is followed by a major event or change in your environment, it's easy to believe the star is controlling things here on Earth. This was how the ancient Egyptians saw it. For example, 5000 years ago the bright star Sirius would rise in the dawn sky, and within a few days the Nile would flood. The Egyptians assumed Sirius was causing the flood.

Many, if not all, religions are based on what we would call 'astrology': prophets would observe the stars and the complex motions of the planets, and attempt to forecast not only seasonal events but also the destiny of humans. The star of Bethlehem, believed to announce the birth of a saviour, is just one example.

When the first towns emerged – the oldest being Jericho in the Jordan Valley about 9000 years ago – it became essential, for organisation and planning purposes, to have an accurate calendar. Structures such as Stonehenge in England are believed to have allowed people to more accurately follow the cycles of sun and moon, and so devise calendars.

Civilisation arose independently in five different locations around the world, each at a different time. Every culture in the world, with the exception of those that remained as hunter-gatherers, can trace their cultural and spiritual origins back to one of these five great traditions. Stone circles appeared on

Stonehenge, on England's Salisbury Plain, was built around the same time as the Great Pyramids of Egypt, although construction spanned several centuries. The giant stones have so many astronomical alignments it is believed the ancient Britons who built the structure may have designed it to track the movements of the sun and moon, and therefore use it as an elementary calendar. SACHA HALL

the scene just prior to the rise of civilisation in their particular area. Thus, they were largely built in the late Stone Age and at the beginning of the Bronze Age. Stonehenge on Salisbury Plains was built at roughly the same time as the Great Pyramids in Egypt – around 4500 years ago. And Stonehenge is not the only stone circle: there are literally thousands around the world. Right across Europe, throughout Africa, across Asia and right here in Polynesia, people were building stone circles. (For more on these fascinating places, have a look at *Stonehenge Aotearoa – The Complete Guide*, see page 119.)

Why were monuments such as Stonehenge built? Nobody can really answer that question because the people who built Stonehenge left no records. We can only speculate from the graves we have found, and by considering the science that may have lain behind the construction. When you begin to study Stonehenge, you find so many astronomical alignments they cannot be there by pure accident. It is obvious that these people were closely observing the cycles of the sun and the moon. From the movement of the sun, its ever-changing rise and set positions, they would have been able to foretell the changing of the seasons. This, along with the cycle of the moon, would have given them their calendar.

THIS BRINGS ME TO STONEHENGE AOTEAROA. Some people have queried why anyone would want to build a version of Stonehenge on the other side of the world, and the sense of it. I even read on one website that, because we were building a Stonehenge in the southern hemisphere, aliens wouldn't know which side of the world to go to when they arrived. However, I find it hard to imagine that aliens who have conquered space would need stone circles for navigation.

In reality, stone circles represent, among other things, fundamental astronomy, and when people come to Stonehenge Aotearoa they suddenly begin to understand celestial mechanics. The henge stones and lintels form windows, which frame the rising and setting positions of seasonal stars, and track the seasonal variation in the rising and setting position of the sun. Additional stones mark the precise moment of the equinoxes and solstices: spring, autumn, midsummer and midwinter.

Stonehenge Aotearoa is not a replica of the original Stonehenge. It is a working model designed for its precise location on Earth, and combines many ancient technologies, including Celtic, Babylonian, Egyptian and Polynesian. Whether you're young or old, if you stand on the henge you can rediscover the knowledge of your ancestors. You will see things happening in the sky just the way they did.

For example, near the centre of the henge is a large obelisk. The shadow this casts gives us a calendar, right down to the actual day. It also shows us, from an Earth-centred perspective, where the sun is, relative to the distant stars. Thousands of years ago the Egyptians, using nothing more than shadows cast by obelisks, not only determined that the Earth was round, but calculated its circumference to a high degree of accuracy.

The henge is also a Polynesian star compass; when you stand in the centre you can see how Polynesian peoples used stars to navigate across the vast Pacific Ocean.

Today, we tend to use latitude and longitude for navigation: in other words, positions are marked relative to the whole Earth. Polynesians didn't do this. They had a home-reference system. Before a boat left an island, its course would be worked out through the stars, and beacon points set up. When the boat sailed, the sailors would align its course with these beacons. In this way, the vessel could set sail before the stars were up for them to navigate by: they could see the beacons until they went below the horizon. Because of this, we believe that many of the stone circles in Polynesia were not to do so much with seasonal changes as with navigation. These seafaring people could work out exactly where they were in the ocean by the rise and set of particular stars, a process they called *rua*. There is a beautiful star compass in Hawai'i, drawn on a big stone, which shows the rise and set positions of stars.

Building Stonehenge Aotearoa has been a big learning curve for all of us who have been involved. Most of all, we have gained a new appreciation for the people who built structures like this thousands of years ago. In order to get everything in exactly the right position, we spent over a thousand hours surveying the

Stonehenge Aotearoa is an open-sky observatory, especially designed for its southern hemisphere location. Inspired by England's famous monument and built to the same scale, it is an aid to understanding early astronomical knowledge, from as far back as Babylonian times. CHRIS PICKING

site. The builders of the original Stonehenge and other stone circles didn't have the wonderful surveying instruments we have today, let alone GPS (Global Positioning System). They had to start from scratch, and even today that's not as easy as you may think. You can have a computer that tells you the sun will rise in a certain spot on a certain day, but will it really? When you see the sun on the horizon, you're not actually seeing the sun at all. You are seeing a mirage, a refracted image of the real sun, which is still below the horizon. As the sun rises up – and it does so at an angle – the mirage image merges with the real sun. Consequently, there will be a difference between the calculated and the observed positions of sunrise.

You can do the theory, but there's nothing like actually doing the experiment, so one day, when we had completed a series of calculations, we went out and erected a post. If our calculations were correct, the sun would rise over this post the following morning.

We gathered beneath the stars in the morning twilight. Eventually, the top of the sun's disk cut the horizon, and sunlight swept across the paddocks. Captured by the morning dew, everything was bathed in golden light. The birds started to sing. For scientists and non-scientists alike, the feeling was awesome: we were standing out there in nature with our ancestors. The sun came up and sat on top of the post. We knew at that moment that we had our calculations right.

***Richard Hall*** *is co-founder of the renowned Stonehenge Aotearoa Astronomy Centre in Wairarapa, New Zealand, an inspirational public speaker, a producer of audio-visual and planetarium shows, and author of the bestseller* How to Gaze at the Southern Stars *and the coming book* Zodiac: Remarkable Stars on the Path of the Sun. *He is a member of the Royal Society of New Zealand, and of its standing committee on astronomy.*

## A Brilliant Civilisation

*—Peter Adds*

WHEN CAPTAIN COOK was in Tahiti making his observations about the transit of Venus, he was among people who were also profoundly good navigators and sailors, a fact he was slow to recognise. When he and the other European explorers arrived in the Pacific they were bemused, frustrated and even agitated to discover that other peoples had beaten them to these islands. They began wondering how people who appeared to be nothing more than savages had been able to cross the world's biggest ocean, find the islands and settle them. These people didn't have compasses, maps or sextants. Nor did they appear to have very good boats. How on earth had they been able to do it?

This question would become a major source of fascination for Europeans for a long time afterwards, and would produce much scholarship. One of the products of that scholarship has been a renewed interest among Polynesians themselves in the arts of sailing and navigation, and an immense sense of pride in the skills their ancestors must have possessed. Today, major efforts are being made across Polynesia, including here in New Zealand, to find and rekindle these arts.

Many of the early theories about the ability of Polynesians to sail and navigate were based on the assumption that, as a race, Polynesians must have degenerated from some former better, more brilliant civilisation. They had to have been good at these arts at one time, the theory went, because they had got to the islands, but after that, in the isolation, they had regressed into the people witnessed by the early European explorers.

This idea of degeneration became very influential, and it took a long time of observing Polynesians and their marine traditions for theorists and scholars to click on to the fact that these people were inherently good at sailing and navigating and building boats. By the middle of the nineteenth century, however, scholars were starting to take the marine capabilities of Polynesians more seriously. The Polynesians' aural traditions talked about great voyages of discovery, great navigators and great explorers, and some ethnographers started to believe there might be something in these stories. Maybe Polynesians did have a marine technology, a body of knowledge that allowed them to sail to distant places?

By the middle of the twentieth century, an orthodoxy had developed about these ideas. It was, however, an orthodoxy based mostly on assumptions – produced by people such as Te Rangi Hiroa (Sir Peter Buck), Percy Smith, and other ethnographers. By the 1940s this orthodoxy had become very persuasive. It went something like this: Polynesians had originally come from the west, from somewhere in Asia, and had sailed into the Pacific against the winds and the currents. This idea was popularised most importantly by Sir Peter Buck in 1938, when he published a book in the United States called *Vikings of the Sunrise*. In it he presented a romanticised idea of Polynesians' sailing capabilities, as well as his theories about their origin as a race – some of which were, probably rather accidentally, not too far from what we know today to be possibly the case.

The orthodoxy also held that Polynesians had large seaworthy canoes, that they could navigate over very large distances whenever and wherever they felt like it, and that their voyages of settlement and exploration were intentional, and the primary means by which Polynesia had been settled. Now, there was a slight problem with this that went unnoticed at the time, which was that all the people putting forward the theory were what we might describe as armchair theorists. They were not sailors, had not been to sea themselves, and did not really have any expertise in sailing matters generally, much less in Polynesian types of sailing.

Nevertheless, the orthodoxy was well entrenched in New Zealand folklore and mythology until the mid 1940s, when some people came along who weren't entirely taken with the theory and started to challenge it. The first was the Norwegian adventurer Thor Heyerdahl. In the Pacific, the winds and currents for the most part go from *east to west*, and we had begun to learn from archaeology that the direction of settlement in Polynesia had been from *west to east*. Heyerdahl, who had an archaeological background of sorts, didn't think Polynesians were clever enough, or had the necessary technical expertise, to have sailed against the wind and currents.

Heyerdahl also knew that the kumara, the sweet potato that was the staple diet of many Polynesian peoples, was a South American cultigen, not a South-east Asian one. He put two and two together, and came up with the idea that the Polynesians had come originally from South America.

In 1947, Heyerdahl set out to test that idea by constructing a boat based on ancient South American Indian designs. That boat, the *Kon-Tiki,* was essentially a raft with a large structure on top, one mast, and a couple of sails attached to the mast. It was built largely of local balsa-wood, bundled and tied together, and designed, among other things, to allow the water to flow through. However, when Heyerdahl and his crew attempted to

Thor Heyerdahl, a Norwegian adventurer, set out in 1947 on a raft, the *Kon-Tiki*, to prove that Polynesian peoples came originally from South America. Although the demonstration failed, Heyerdahl became an international celebrity, and his book on the voyage was translated into around 50 languages. Today, genetic evidence has shown that Polynesians are likely to have come from South-east Asia.

sail out of Callao Harbour in Peru there was a problem: a little thing called the Humboldt Current, which runs swiftly off the western coast of South America in a northerly direction. Caught up in the current, the *Kon-Tiki* was unable to get out of it. It was being swept anywhere but Polynesia. In the end, Heyerdahl had to hire a tugboat to tow the craft 50 miles out to sea across the Humboldt Current.

You might have thought this would negate the experiment, but Heyerdahl continued anyway. The *Kon-Tiki*, being constructed of balsa-wood, soon became waterlogged. And as it sailed in a westerly direction into Polynesia, it became heavier and heavier, and more susceptible to currents, and therefore more difficult to steer. Heyerdahl and his crew did make it to Polynesia eventually, sailing with the wind and going with the currents, but unfortunately they had to pass a whole lot of islands they sighted along the way: the *Kon-Tiki* had become so waterlogged it couldn't be steered.

Finally, about 101 days after they had set out from Peru, the group crash-landed on Raroia, a small island in the Tuamotus. Heyerdahl had proved that if you get towed by a tugboat across the Humboldt Current and are lucky enough, you can crash-land on a small tropical island in Polynesia. The scholars of Polynesian history and people who were interested in sailing and navigation did not accept Heyerdahl's theory, but the experiment had wide public appeal. Heyerdahl's book on the voyage sold millions of copies and was translated into around 50 languages. There were television programmes and an Academy Award-winning documentary.

A few years later another doubter came along, this time a New Zealand historian called Andrew Sharp. Sharp was an armchair theorist who, like Heyerdahl, did not believe the early Polynesians would have had the technical ability to sail against the winds and currents. Nor did he think their canoes were good enough, or their knowledge comprehensive enough to navigate between islands over large distances. In his view, therefore, the settlement of Polynesia had been one big accident, a theory he expounded in his 1956 book *Ancient Voyagers in Polynesia*. He argued that people fishing at sea had probably got lost in storms and ended up accidentally arriving on islands, where they subsequently settled. And incidentally, virtually every piece of Pacific real estate that can sustain human life was settled at some time by Polynesians.

But there was also a problem with Sharp's theory, which was that in Polynesia fishing tends to be done solely by men. And if you are going to subscribe to the idea that the settlement of Polynesia was an accident, you have to have some women in there. This was largely overlooked by Sharp.

What Heyerdahl and Sharp did, though, was get people thinking about what had actually happened. What was needed was some hard data. Enter another New Zealander, a man by the name of David Lewis. Lewis was an explorer, sailor and doctor who had spent some time sailing around the Pacific in his own boat. In Micronesia he had met up with some local sailors and learned – especially from Mau Piailug, a well-known man from the island of Satawal in the Caroline Islands – some of the lost arts of navigation. These techniques included navigating by star positions; reading land signs – signs that indicate the proximity of land when at sea in the open ocean; understanding the directional qualities of currents, swells, and the phenomenon called luminescence that appears in the sea at night; and being guided by wind directions and cloud formations. It was a very comprehensive body of knowledge.

In the early 1970s, using these techniques, David Lewis successfully sailed his trimaran from Tahiti to New Zealand, thereby proving that Polynesians had the technical expertise to undertake the amazing voyages of colonisation we know must have happened, and that they could have not only undertaken voyages to new homelands, but turned around and sailed home again relatively easily had they wanted to.

Just how miraculous those voyages were was shown in an ingenious computer simulation carried out in England in 1972 by three meteorologists, Michael Levison, Gerard Ward and John

Webb. These men gathered a vast amount of Pacific weather information, including currents, winds and storm conditions, put it together in their supercomputer with data on canoe speeds at different wind speeds and ran a series of over 100,000 simulated voyages. Their aim was to find out what would happen to canoes leaving various islands if they were left to drift.

They found it would be virtually impossible to drift accidentally to many of the islands settled by Polynesians. For example, if you drifted in a canoe from Samoa you would have only one chance in 700 of ending up in the Marquesas. Yet we know that people did make such journeys across the Pacific in an easterly direction. The obvious conclusion was that the voyaging was not accidental but intentional: the seafarers must have deliberately sailed against the wind and currents to get to the places where they eventually settled.

This computer simulation and David Lewis's work were compounded in 1976, when a replica Polynesian voyaging canoe was built in Hawai'i. The *Hokule'a* was a double-hulled canoe modelled on ancient Polynesian designs, but made in fibreglass rather than timber. After it was sailed for several years around the Hawai'ian islands, a voyage from Hawai'i to Tahiti was planned. Traditional Polynesian navigational techniques would be used, with Mau Piailug and David Lewis as crew members.

The voyage of around 3000 kilometres lasted just under a month. The crew sailed without maps, compasses or sextant. The only means of navigation was Mau Piailug. The Micronesian had never sailed the route before, but he knew from knowledge handed down to him by his grandfather the route that should be taken to make landfall in Tahiti. On the journey they were followed by a *National Geographic* support boat, so Piailug's route was able to be scientifically checked. It turned out that his daily calculations about where they were in the Pacific Ocean were never more than 40 miles out. When you think of all the forces acting on the canoe – the different wind patterns, the fact the night sky was often clouded over so Piailug couldn't see the star positions and had to rely on currents and swells – it was an astounding feat. When the canoe arrived in Tahiti, the Tahitians declared a national holiday to celebrate the reawakening of the lost art of navigating.

Since the construction of the *Hokule'a*, a number of other ethnographic replica canoes have been built in the Pacific. In Northland, New Zealand, a man called Hek Busby made *Te Aurere*, a double-hulled ocean voyaging canoe that has since sailed around the Pacific a number of times. Another New Zealander, Greg Whakataka-Brightwell from Ngati Toa on the east coast, built a canoe, *Hawaiki-nui*, in Tahiti of New Zealand totara. Sir Tom Davis, the former premier of the Cook Islands, built a double-hulled voyaging canoe, *Te Au-o-Tonga*, and formed the Cook Island Voyaging Society, with plans to sail to South America and retrace the voyage where Polynesians picked up the kumara and brought them back into the Pacific. All these canoes have been involved in retracing the steps of our ancestors, and reinvigorating Polynesians' awareness of their maritime legacy.

Today, then, we know essentially where the Polynesians came from and we know roughly when. We know they sailed the biggest ocean in the world relatively easily: the Pacific must have been virtually buzzing with canoes at some time in the ancient past. The big question remains why. Why did Polynesians make these amazing voyages into the Pacific? And why did they

The Hokule'a, a double-hulled canoe which replicates early Pacific seafaring craft, completed an astonishing voyage from Hawai'i to Tahiti in 1976, using only early Polynesian navigational methods. Since then it has completed a further six Pacific voyages, and as this book went to press it was undertaking an epic voyage from Hawai'i to Micronesia and Japan. MONTE COSTA PHOTOGRAPHY

eventually settle all of the islands of Polynesia that are capable of sustaining human life?

One theory centres on wanderlust: Polynesians simply wanted to see what was over the horizon so they went and had a look. This idea of an innate curiosity and a tremendous desire to explore is bound to be right at one level. There is no doubt that early Polynesians had the technical capabilities to sail easily to virtually anywhere they wanted, so what was to stop them going and having a look and maybe finding a new island to live on? That must have been part of that story.

Another associated theory is that of overpopulation: Polynesian communities on small tropical islands quickly grew to a point where pressure on resources was too great, and people were forced to leave and find new homelands. Presumably, they travelled for the most part in an easterly direction, although we do know that some went west again, back into Melanesia, to occupy the islands we refer to today as the Polynesian outliers.

Demographers, however, tell us that people appear to have left for new islands before population levels on their home islands reached critical levels. If the demographers are right, overpopulation may not be the critical factor. It is bound to be part of the story but it can't be the only explanation.

Another very popular theory, sometimes expressed in the anthropological literature, is that migration was spurred by sibling rivalry – fighting between brothers and sisters. Polynesian societies are very hierarchical. At the top of the social pyramid is the chief, and at the bottom are the commoners and slaves. The various layers in between depend on which Polynesian society and culture you're talking about, but the whole structure is inherently unstable. If you take a society like that and place it on a small tropical island where the population is going to grow rapidly, some anthropologists argue that there is only one possible outcome and that is tension.

This is supported by a large number of stories in Polynesian mythology in which one way out for people stuck in the middle of the social order but politically ambitious is to build a canoe with your family, head off out to sea, find a new island, colonise it and become the new chief. If you're the younger brother of the chief, there is another alternative: you kill your older brother and assume his position. Both sets of stories are found.

The truth, though, is we don't know why Polynesia was settled the way it was. There are big gaps in our knowledge.

Incidentally, some of the places where those early Polynesians settled were pretty difficult places to find. Easter Island, for example, is the most remote piece of dirt on the planet. If you're on Easter Island you're further away from any other piece of land than you would be anywhere else in the world. What were those early voyagers doing way out there?

It's a similar story for New Zealand. We know, of course, that New Zealand was settled by people from East Polynesia. There are arguments about precisely when this happened, but probably it was around about 1200 to 1300 AD. Because of the ocean currents, you can't get to New Zealand accidentally from East Polynesia: you have to be sailing in this direction quite deliberately. Semantically, you can call New Zealand's discovery an 'accident' because people did not know it was here in the first place. But it was clearly found by someone sailing this way deliberately. And chances are that whoever that person was – Kupe, Toi or someone else – they probably turned around, sailed home and told all their relatives about this wonderful new land in the South Pacific, and that started a series of migrations.

The best we can say at the moment is that these people came from East Polynesia. We can't hone in on any particular island homeland – mainly because the material culture, the artifacts, of that time from the various islands of East Polynesia are so similar it is impossible to distinguish between them. But in the not-too-distant future, through the study of human DNA, we will undoubtedly learn a lot more about those migrations. We will know precisely who the nearest relatives, genetically speaking, are of Maori people. Kia ora.

***Peter Adds,*** *Te Ati Awa, is head of Tumuaki, the School of Maori Studies at Victoria University, Wellington, and a member of the boards of the Historic Places Trust and the Maori Heritage Council. He is a negotiator for the treaty claim of Te Ati Awa ki Taranaki.*

## To the Farthest Ends of the Earth

*—Duncan Steel*

THE PREDICTIONS AND OBSERVATIONS of Venus crossing the face of the sun were, over two centuries, important in the development of science and navigation, and linked to the histories of New Zealand and Australia. But they were also amazing stories of human endeavour, from sled-rides on frozen rivers to endurance of the sweaty, disease-laden heat of the tropics, from attempts by Siberian peasants to lynch an astronomer to the willingness of Tahitian maidens to sell themselves to British sailors for a ship's nail or two. And so we begin.

In antiquity, it was thought that the stars and planets were just above our heads. All celestial objects were believed to revolve around the Earth, and the distances between them were imagined to be relatively small. Modern astronomy began in the early sixteenth century, when Nicolaus Copernicus described a system in which the planets, including Earth, orbited around the sun. Even though this idea had been suggested 2000 years earlier by the Greek mathematician and astronomer Aristarchus, it was seen as revolutionary, and much argued about. In particular, there were religious problems: the established church did not like the relegation of Earth to the status of just an ordinary planet.

Even if you accepted the Copernican model of the solar system, there was no idea of the real distances involved. How far was it, for example, from the sun to the Earth? This is a fundamental parameter in astronomy, and termed the Astronomical Unit, or AU. The great Danish astronomer Tycho Brahe, born three years after Copernicus died, thought the distance was about eight million kilometres, or five million miles. Actually, it is almost 20 times that far.

Around 1600 there were two major steps forward in astronomy. First, the telescope was invented. When Galileo turned his rudimentary telescope toward the heavens, he saw things no one had seen before. People had imagined that the heavens were essentially perfect, but when Galileo looked at the surface of the moon he saw craters and mountains. In short, the moon was not perfect. When he looked at Jupiter, he found it was accompanied by four apparent bright stars, which moved. Later these were recognised to be natural satellites of the planet; today we term them the Galilean moons. Again, this was a sign of the newly discovered imperfection of the heavens.

The other great step forward came in 1609–19, when the German mathematician Johannes Kepler introduced his three laws of planetary motion – the basic mathematics that govern the way in which objects orbit the sun. Using these laws, Kepler went on to predict in the late 1620s that both Venus and Mercury would cross the face of the sun in 1631. Kepler died in 1630, so he was not around to see these events for himself. Indeed, no one observed the 1631 transit of Venus, because to do so you needed to be in the Pacific region, and there was no time for the news to be carried to, say, the Spanish settlements in Mexico. However, the Mercury transit in November of that year was observed by astronomers in Paris and elsewhere in Europe. This confirmed the correctness of Kepler's laws and analysis. It was a triumph of prediction.

Kepler, though, had missed something. He thought no further transits of Venus would occur until 1761. This error was due to his lack of knowledge of the distance to the sun: his value was seven times too low. Transits of Venus, we now know, occur in pairs eight years apart. The reason for this is that one Venus year equals 8/13ths of an Earth year. That is, in the time it takes the Earth to orbit the sun eight times, Venus does 13 laps.

We also know that transits of Venus occur with spacings of 121.5 years, then eight years, then 105.5 years, and then eight years again. So a pair will occur in June, eight years apart, then more than a century will elapse before there is another pair. These will also occur eight years apart, but in December.

This is a good example of what we might call 'the clockwork of the heavens', a regularity first figured out in the late 1630s by Jeremiah Horrocks, a remarkable young man from Liverpool. We know comparatively little about Horrocks. We believe he was born in 1618. We know that between the ages of 14 and 17 he studied theology at Cambridge University, but that his real interest was in astronomy. We know that by 1639 he had left Cambridge and was working as a tutor to children of gentry in the Lancashire village of Much Hoole, just south of the town of Preston. He was also spending time assisting in the local church. We know for sure that he died in 1641 – at only 22 years old – and that his works were not published until decades later.

Horrocks was an early advocate of Kepler's laws. In fact, he was almost unique in Britain in believing them to be correct and recognising their significance. He also did important new astronomical work himself. He was the first to show that the orbit of the moon around the Earth is elliptical. He made occultation observations – that is, observations of the moon crossing the paths of stars in the sky – and noticed that the stars suddenly disappeared when they went behind the edge of the moon. He interpreted this as showing, first, that the moon was airless – that is, it lacked an atmosphere – and secondly that the stars must be point sources, which indicated to him that they must be at great distances.

Late in 1639, Horrocks realised that Kepler had been wrong in saying that no further transits of Venus would occur until 1761, and that a transit would in fact be occurring in less than a month. He had time only to write to his brother Jonah, who lived in Liverpool, and his friend William Crabtree, who lived in Salford, a city 30 miles away in Greater Manchester, and alert them to the coming event.

On the day it was mostly cloudy in both Liverpool and Manchester. In gaps between the clouds, Crabtree managed to see Venus on the face of the sun. However, he had no time or opportunity to get accurate measurements. Horrocks, meanwhile, to the north in Much Hoole, saw the transit begin about 30 minutes before sunset. From his measurements he was able to show that the diameter of Venus was about one twenty-fifth of the apparent diameter of the sun. This was far smaller than Kepler had thought.

Previously, it had been imagined that the sun was much closer than it is, and it had been this, in essence, that had caused Kepler to miss the 1639 transit in his predictions. From his observations, Horrocks now calculated that the distance to the sun must be about 90 million kilometres, or 60 million miles. Although this was actually only two-thirds of the real distance, it was much, much further than had previously been suspected, and was frightening new information for people to take on board, showing, as it did, the huge distances between the planets.

Horrocks corresponded with Crabtree for several years. The two did much joint work, but never met. Horrocks was due to go to meet his collaborator on 4 January 1641 to discuss their astronomical research. However, on 3 January he mysteriously dropped dead. Only three years later Crabtree also died, aged 34.

After Horrocks's death, most of his voluminous writings were lost, many by burning. This had religious overtones. Lancashire was the Northern Ireland of the era, with great animosity between the Protestants and the dominant Catholics. Horrocks himself was a Puritan. The knowledge he was introducing was dangerous, and it is believed much of his writing was burnt by the Catholics. He was then largely forgotten, and it was not until the Victorian era that he was rediscovered. In 1874 a plaque was erected to him in Westminster Abbey, right opposite that of Isaac Newton. This was fitting: had Horrocks lived, it is likely he would have anticipated much of Newton's research work, and Newton certainly knew of Horrocks's writings.

If you go back to Lancashire now, you will find that in the church in Toxteth, the area of Liverpool where Horrocks was born, there is a plaque commemorating him. And at Much Hoole, where he observed the transit, there are not only plaques but also a chapel named for him, and stained-glass windows showing him observing the transit.

In Manchester, Crabtree is remembered. If you go to the town hall you will see a set of murals celebrating the history of the town, and in one of them Crabtree is depicted observing the transit. However, considerable artistic licence has been used: Crabtree looks about 70, whereas we know he died when less than half that age. The technique he is using, though, is exactly

The church in the village of Much Hoole in Lancashire, where Jeremiah Horrocks lived at the time of the 1639 transit of Venus, contains memorials to him, including this stained-glass window. He wrote of the event: 'About fifteen minutes past three in the afternoon ... the clouds, as if by divine interposition, were entirely dispersed ... I then beheld a most agreeable spectacle, the object of my sanguine wishes, a spot of unusual magnitude and of a perfectly circular shape, which had already fully centred on the sun's disc on the left ... Not doubting that this was really the shadow of the planet, I immediately applied myself sedulously to observe it.' DUNCAN STEEL

the way we do it today: rather than looking directly at the sun, he is using a telescope to project its image on to a screen. You should *never* look directly at the sun, especially not through a telescope or pair of binoculars.

Transits of Mercury occur more frequently than those of Venus, with spacings of three, seven and 13 years. Typically, in any century it will be possible to see 12, 13 or maybe 14 such transits. After 1640 several transits of Mercury were seen, and in 1651 another Lancashire man, Jeremiah Shakerley, travelled all the way to India to observe one.

Jeremiah Horrocks, having calculated that a transit of Venus would occur on 24 November 1639, had time only to alert one other astronomer to the impending event: his friend William Crabtree, who lived in Salford, part of Manchester. This plaque, unveiled in June 2004 to mark the transit of Venus that year, commemorates Crabtree's historic observation. DUNCAN STEEL

IN 1663, A SCOTSMAN, James Gregory, suggested how a more accurate measure of the distance to the sun could be obtained from observing a transit of Venus. Gregory, who was born in 1638, the year before Horrocks observed his transit, made a number of seminal contributions to astronomy. He invented the first practical type of reflecting telescope (that is, a telescope using curved mirrors rather than the lenses that Galileo employed when he first turned a telescope towards the heavens in 1608). This – the Gregorian telescope – was to remain the standard form of reflector until the nineteenth century. Gregory also invented the diffraction grating, which is still the device that astronomers most often use to split starlight into its component colours and wavelengths, and he did this just a year after

Newton conducted his famous experiments on the composition of light using a prism. What Gregory used as a 'diffraction grating' was simply a bird's feather.

The method Horrocks had used to estimate the separation of the Earth from the sun was crude, but the best he could do as a solo observer. Gregory realised that a far better measurement might be obtained if at least two well-separated observers were involved, and his concept was later taken up by others, as we will see.

At this stage I need to introduce the name of one of the most famous astronomers of all time – Edmond Halley. Halley knew that the comet that now bears his name would return in about 1758, long after his death. When he died in 1742, he left a note saying he hoped history would record that it was an Englishman who had predicted the comet's return.

He made similar predictions about the transits of Venus in 1761 and 1769. He, too, realised that by observing a transit of Venus from two widely spaced locations on the Earth's surface, it might be possible to determine the distance from the Earth to the sun with considerable accuracy. He thought the sun was 111 million kilometres, or 69 million miles, away, but knew his value was inaccurate and wanted to find a way to get the distance measured with more precision.

There were two reasons for wanting to make such a measurement. First, astronomers needed to know the scale of the solar system – a fundamental cosmic distance – for reasons of pure science: knowledge for its own sake. There was, however, another reason. Halley had a theory on the motion of the moon, and he knew that in order to improve this theory he must measure the distance to the sun accurately. He could then calculate the perturbations of the moon caused by the sun and the other planets. This was not simply for pure knowledge: Halley's eventual aim was to be able to produce a table of positions of the moon at arbitrary future dates. Seafarers could then take this table on their voyages, and by observing the position of the moon would be able to tell the time. And from the time they would be able to determine their longitude, which was a much desired quest in the eighteenth century.

How did Halley get this idea? In 1677, at the age of 21, he had observed the transit of Mercury from Saint Helena, a little island in the south Atlantic, and found that his observed time for the transit agreed with his prior prediction within an accuracy of one second. Fourteen years later he suggested to the Royal Society that a transit of Venus (rather than of Mercury) would be useful for determining the solar distance, and urged the society to carry out precision observations in 1761 and 1769.

The basis of Gregory and Halley's method was what we call 'parallax'. Imagine that you gaze at something in the distance – say, a group of trees – and hold up a finger at arm's length. If you look at it first with one eye and then with the other, blinking your eyes in turn, your finger will appear to jump sideways compared to that distant background. This is due to your eyes being separated.

In principle, you could measure the length of your arm this way. Of course it would be easier to use a tape measure, but we can't use a tape measure to measure the distance from the Earth to Venus, or from the Earth to the sun. If, however, astronomers were to view a transit of Venus from two well-separated locations on the Earth's surface, they would, Halley knew, see the planet taking different paths across the face of the sun. Venus

A Scottish mathematician and astronomer, James Gregory, first conceived of the idea that the distance from the Earth to the sun could be precisely measured by observing and timing the transit of Venus from different locations, a technique known as parallax. His idea was taken up by English astronomer Edmond Halley (pictured), who urged the Royal Society to make precise observations of the 1761 and 1769 transits. His expectation was that more accurate navigational tables would result. This portrait of Halley, who became Astronomer Royal, was painted circa 1687, when he was in his early thirties. He is most famous for the comet that bears his name.

would, like your finger, be moving compared to the background. This is the technique of parallax.

The two paths taken in such a case are straight lines, but they are *different* straight lines, separated by a small distance. In order to calculate that distance accurately, we need to measure the length of the lines across the apparent circular shape of the sun. And the most precise way to do that is not to try and measure their lengths directly, but rather to measure the time, or duration, of the transit. Since it is possible to calculate how fast Venus is moving, the duration of the transit, as seen from different positions on the Earth and accurately measured to a matter of seconds, will allow us to know very precisely the particular course taken by Venus across the face of the sun.

Mercury is too far from the Earth for this purpose, but Venus is at an appropriate distance. Both Gregory and then Halley realised this. Halley wrote, 'By this means, the sun's parallax may be discovered to within its five-hundredth part. That is, an accuracy of 0.2 percent.' In fact, as we shall see, Halley was overoptimistic because he based his estimates only on his observation of Mercury in 1677, and it happens that there are more pitfalls to measuring a transit of Venus than that of Mercury. This, though, could not be known in Halley's era.

As the eighteenth century progressed, many European nations made plans to observe the transit. This was the space race of the era, and there was much national pride at stake. The exercise was hugely expensive, in terms both of money and men, but the essential problem being tackled was of huge importance: navigation and finding one's longitude.

In 1761, then, many teams were sent out to observe the transit. The Astronomer Royal, Nevil Maskelyne, went to Saint Helena. An Austrian Jesuit astronomer, Father Maximilian Hell, went in the opposite direction, to Lapland in Norway, and got good observations from there. Other astronomers went to Newfoundland, where they were pestered by mosquitoes and other flying bugs.

Another team was Charles Mason and Jeremiah Dixon, whose names would later become well-known because of the Mason-Dixon line dividing the United States into north and south, a line which they surveyed in the mid 1760s. In 1761 the pair were contracted by the Royal Society of London to sail to Indonesia and observe the transit. The problem was that the Seven Years' War between Britain and France was in full swing, and no sooner had Mason and Dixon's ship left Portsmouth Harbour than it was attacked by a French frigate. Eleven sailors were killed and many more injured, and the ship, severely damaged, put back into Plymouth. Mason and Dixon then petitioned the Royal Society to relieve them of their charge. When the society insisted they continue, they sailed down through the Atlantic but gave up their voyage when they got to South Africa, where they were able to observe only the exit of Venus from the face of the sun.

The French also sent out many observing teams. One, under an abbot, Alexandre-Gui Pingré, went to the island of Rodrigues in the Indian Ocean. Another, led by a 41-year-old aristocrat called Jean-Baptiste Chappe d'Auteroche, went to Tobolsk in Siberia, 5000 kilometres east of Paris. Chappe d'Auteroche started out eight months before the transit. His horse-drawn sled just made it across the frozen Volga River before the ice pack broke up. His party then had to travel through a sea of gooey mud for thousands of kilometres, slowing him down to such an extent that he got to Siberia only days before the transit was to occur. There he had to be protected by armed Cossacks because local peasants thought the severe spring floods were caused by this foreigner messing with the sun. Despite such difficulties he managed to obtain good timings of the transit, but it then took him 18 months to get back to Paris.

Another French aristocrat who went to observe the transit was Guillaume Joseph Hyacinthe Jean-Baptiste Le Gentil de la

Galaisière, who departed France 15 months early, aiming to get to Pondicherry, a French possession in the east of India. When he reached Mauritius, he learned the British had Pondicherry under siege and the Indian Ocean was full of hostile ships. He eventually got on to a troop ship bound for Pondicherry, but when the ship's captain heard that the town had fallen to the British four months earlier he turned back to Mauritius. The transit occurred under clear skies, but with Le Gentil still at sea on his ship he couldn't make any useful observations.

Le Gentil then decided to stay in the Indian Ocean region and wait for the next transit eight years later. However, he didn't waste his time, making various natural history studies in Mauritius and Madagascar, investigating the islands' botany, zoology, geology and anthropology, before finally deciding to go to Manila, the capital of the Philippines, for the 1769 transit. We will read more about Le Gentil's transit adventures below.

WHEN GATHERED TOGETHER and analysed by mathematicians, the 1761 transit observations proved unsatisfactory. However, it was confidently expected that, with this experience in hand, the 1769 transit, the last for more than a century, would produce better results. By 1765 Thomas Hornsby, the Oxford professor of astronomy, was urging European nations to prepare their expeditions. 'Posterity,' he wrote, 'must reflect with infinite regret their negligence or remissness; because the loss cannot be repaired by the united efforts of industry, genius or power.' In other words, miss this chance and that is it until 1874.

The Royal Society of London successfully petitioned George III for funding, and the king had an observatory erected at Richmond upon Thames, southwest of London, so he could observe the transit himself. Father Maximilian Hell went to Norway again. Charles Mason, now without Jeremiah Dixon, was also sent to Norway. William Wales, another British astronomer, having asked to be sent somewhere warm, was dispatched to Fort Churchill on Canada's Hudson Bay, where polar bears kept him company, along with mosquitoes, flies and horseflies so big he would claim they tore out chunks of his flesh. He reported that even though his cabin had walls three feet thick, his bed became iced up. Once the thaw came, he was pleased to be able to get out hunting geese and moose, and catching salmon. In spite of his pestering by insects, he got good transit timings. Later, in 1772–75, Wales would be on James Cook's second voyage to the Pacific.

The Frenchman Jean-Baptiste Chappe d'Auteroche wanted to go to the Philippines, but Spain, whose colony it was, didn't trust him. Instead, with a small Spanish military escort he went to less sensitive territory, the tip of Baja California in Mexico. The journey was arduous. The party sailed across the Atlantic, through the Caribbean and then travelled by land across Mexico. When they reached the Pacific coast, they built wooden ships to get them to Baja California.

Unfortunately, when they finally arrived a typhoid epidemic was sweeping the area, and many of the party died. Chappe d'Auteroche observed the transit on 3 June, but then himself became ill. Two weeks later he managed to rise from his sickbed to observe the lunar eclipse that would enable him to get the coordinates – latitude and longitude – of his observing location. His illness then worsened, and a month later he was dead. Most of the members of the party succumbed to this illness. Only three survived and set off for home, and two of these died on the way. The single survivor, an engineer named Pauly, eventually reached Paris with Chappe's data in 1770.

Was Chappe the most unlucky expeditioner? That is a difficult question to answer. He got good data but he died, whereas our old friend Guillaume Le Gentil had quite the opposite experience. After his scientific adventures in Mauritius and Madagascar, Le Gentil was ready to try again for the transit of 1769. The good news was that the war with Britain was over, and so he was able to travel more easily and safely. In 1766, three years early, he left Mauritius bound for Manila on a Spanish ship. He was making sure he got there on time. However, when he reached Manila several months later, he found the governor distrusted foreigners and thought he was a spy. Rather than face arrest, our hero sailed onwards, first to Portuguese Macau and then to Pondicherry, back in French hands. He was welcomed warmly by the French governor and allowed to build an excellent observatory with which to observe the transit.

After more than a year of preparation, Le Gentil was ready. The weather looked excellent. For weeks before the great event there was not a cloud in the sky. Then, at two in the morning, Le Gentil was awoken by the sound of the wind altering direction. Hastily going outside, he saw a bank of clouds approaching and realized that at 7 a.m., when he was due to observe the end of the transit, the sky would be covered. He returned and lay face down on his bed, unable to raise himself from his despair. It was cloudy all day. He later wrote in his journal, 'I was more than two weeks in a singular dejection and almost did not have the courage to take up my pen to continue my journal and several times it fell from my hands when the moment came to report to France the fate of my operations.'

It was 1770 before he managed to get a passage back to Mauritius, and there was another long wait until eventually he got on to a French ship. After this was almost wrecked off the coast of South Africa, he talked himself on to a Spanish ship setting off for Cadiz in the far south-west of Spain, but that too was almost wrecked. Eventually, in August 1771, he reached Cadiz. By then he had had enough of the sea and decided to walk home.

At last, in October, he crossed the Pyrenees, more than eleven years after leaving France. But his trials and tribulations had not ended. When he got to Paris, he found his home had been ransacked and his relatives, having assumed he had died, had divided up his estate. The Académie des Sciences had also heard he had died and so had given his professorial chair to someone else: Le Gentil was out of a job. His only recourse was simple: he decided to sue them all. However, his luck soon changed. He met and married a wealthy widow and they had a daughter upon whom he doted. Le Gentil's account of his travels and his scientific studies in Mauritius and Madagascar sold well and made him famous. He lived for another 21 years, dying in 1792, just before the terror that was to claim the lives of so many of the French nobility.

The French expeditions were not planned and executed in isolation. Despite wars and rivalry, other nations were involved in coordination. To get as wide a latitude coverage as possible, it was necessary to head south. It is often said that James Cook's expedition was sent to Tahiti because the transit of Venus could not be seen from Europe, but this is wrong; in early June, shortly before the summer solstice, the sun was visible most of the day from far northern latitudes, especially within the Arctic circle, and observing teams went to Lapland and also to Canada.

Tahiti had been mapped by the British only a year or so earlier, and was an obvious choice for Cook's destination. In fact, astronomers planning transit observations, although ignorant of Tahiti's existence, had previously pointed to the area as the ideal location for an expedition. Alexander Dalrymple, a fellow of the Royal Society, was widely favoured to command the expedition. However, the Royal Navy refused to accept him. Its previous experience with civilian commanders had included Edmond Halley decades before – a trip that had ended in near mutiny. It wanted one of its own and chose James Cook.

In 1768 Cook's ship, the *Endeavour*, left Plymouth for the south seas. On board was Charles Green, an astronomer, and two other scientists, Joseph Banks and his Swedish friend Daniel Solander, both botanists. Seven months later the ship reached Matavia Bay in Tahiti. Cook and the scientists set about building their temporary observatories in three separate locations in case there were clouds on the day.

The crews, though, were distracted. The captain of HMS *Dolphin* had had trouble the year before with sailors surreptitiously pulling nails from the ship's decks, almost sinking the ships. Young Tahitian women, not subject to the same moral codes pertaining in Britain, were happy to sell their sexual favours for a single nail. Metal was almost unknown in the islands, and so nails were regarded as a valuable commodity. Knowing this, Cook had had the foresight to take a barrel of spare nails on his ship, but soon even this supply was exhausted and he had to issue strict orders with regard to sailors stealing nails from the boat itself. The episode provides an interesting lesson in economic theory. The nails flooded the market to such an

Cook's voyage from Plymouth, England to Tahiti in the *Endeavour* took seven months. By the time the ship arrived back in England in 1771, after exploring the New Zealand and Australian coastlines, half the crew, including the astronomer Charles Green, would be dead, having contracted fever and dysentery in Batavia, modern-day Jakarta. However, the scientific information the expedition had gathered was so notable that Cook was next sent with two ships, the *Resolution* and the *Adventure*, to explore Antarctica. AUCKLAND ART GALLERY/G-2134-1/1, ALEXANDER TURNBULL LIBRARY

extent that when Cook returned to Tahiti a few years later the going rate for sex had increased to eight or ten nails.

On the day of the transit, the sky was clear and all three observation sites returned transit timings, although with considerable scatter due to what is termed the 'black drop' effect. More on that later. First, let us consider what happened after the transit. Cook had taken with him secret sealed orders, which were not to be opened until after the astronomical event that was the primary aim of the expedition. No one could be sure what these instructions involved. It had been rumoured in London that they would be told to sail the southern Pacific

Ocean looking for the unknown but anticipated great southern continent, *Terra Australis Incognita*.

It turned out that this was exactly what the orders mandated. Sailing westwards, Cook reached New Zealand. The land's presence was known but the islands had never been properly mapped, and Cook spent six months charting the coast, sailing all around it and landing in many different places. Many of the names he gave these places persist today. Banks Peninsula was named for Sir Joseph Banks. Solander got a tiny island southwest of Fiordland. Mercury Bay on the Coromandel Peninsula was so named because Cook and his scientists watched a transit of Mercury there in November 1769. Cook had an almanac indicating the precise time of this transit, which would enable him to deduce the longitude of New Zealand, a measure obviously very useful if others were to follow him.

After New Zealand, Cook and his retinue sailed west across the Tasman Sea and reached Australia at a place they named Botany Bay. From there they continued 3000 kilometres up the east coast until the *Endeavour* was holed on the Great Barrier Reef, forcing them to beach it for some weeks for repairs near modern Cooktown in northern Queensland. Eventually they got to Batavia (modern-day Jakarta), where malaria and dysentery led to the deaths of dozens of his men. In fact, half of the crew, including the astronomer, Green, would die before they reached South Africa two and a half months later.

By the time Cook got back to England in 1771, Thomas Hornsby, the Oxford astronomy professor, had more or less decided what the distance to the sun should be, so his selection of data from Cook's observers was peculiar and biased. On top of this there were three major problems with the data analysis.

The first was the black drop phenomenon mentioned earlier. As Venus entered the solar disc, Cook reported seeing an elongated black ligament, similar to the shape of a water droplet, connecting the dark dot of Venus to the edge of the sun. This was due to sunlight being refracted or deviated in the upper Venusian atmosphere, and other effects to do with the Earth's atmosphere.

As a result, even though observers may have been stood next to each other, there was a scatter of up to half a minute in their timings, whereas Halley's planned technique required a precision better than a second, or two at most. Cook wrote in his diary:

> *This day proved as favourable to our purpose as we could wish. Not a cloud was to be seen the whole day and the air was perfectly clear yet we very distinctly saw a dusky shade around the body of the planet which very much disturbed the times of the contacts. We differed from one another in observing the times of the contacts much more than could be expected.*

The next problem was that the relative positions of the observers spread around the globe were uncertain. It was difficult to measure accurately enough their latitudes and longitudes, and that led to difficulties interpreting the data.

The third problem was purely mathematical: combining multiple observations required methods and techniques that were not then available. (They would finally be developed in the early nineteenth century.)

The next transits of Venus occurred in 1874 and 1882, and were also closely observed. In 1874 Russia alone sent out 26 expeditions, Britain a dozen, the United States eight, France and Germany six each, Italy three and the Netherlands just one. Good geographical coverage was needed, but since the transit was in December there was a better chance of getting clear skies in the southern hemisphere. As a result, the British sent expeditions to Christchurch in New Zealand, a variety of locations in south-eastern Australia, including the newly built Sydney Observatory, and several other places. The Americans sent expeditions to Beijing (then called Peking), Vladivostok and Nagasaki in the northern hemisphere, to Bluff, Queenstown, the Chatham Islands and several other locations in New Zealand, to Hobart in Tasmania, and even to the windswept Kerguelen Island in the Indian Ocean. French expeditions went to places such as New Caledonia and Campbell Island, and a German expedition went to the subantarctic Auckland Islands.

Using the data gathered from these observations, the American astronomer Simon Newcomb calculated a value for the distance to the sun of 149.59 ± 0.31 million kilometres, a precision of one part in 480. This was more or less in line with what Edmond Halley had expected almost two centuries before: that he would be able to determine the solar distance to one part in 500. There was by that time, however, no longer any need for the lunar tables that had been Halley's main motivation. The development of the marine chronometer by John Harrison and others had solved the problem of longitude determination and navigation at sea. Indeed, on his second and third voyages to the Pacific, James Cook had used marine chronometers and sworn by them as the best way to navigate.

One big difference from the eighteenth-century observations of Cook and others and those of the nineteenth century was the advent of photography. There was a particularly interesting footnote to this. Jules Janssen, a Frenchman who observed both the 1874 and 1882 transits, invented a circular glass photographic plate with which, during a transit, he could take up to 60 pictures of the sun – one every second – using a small clockwork mechanism to drive the plate around. Later, Thomas Edison met Janssen and viewed this apparatus, and it inspired him in the development of the movie camera; the whole history of the movies was changed by the fact that Edison saw how to make a movie camera using a technique developed in order to make measurements of a transit of Venus.

In 1882, the transit was visible throughout the Americas, from Canada to Patagonia. Expeditions also went to, among other places, South Africa and the North Island of New Zealand.

In the following years, other techniques were developed by astronomers to measure the distance between the Earth and the sun. For example, the discovery of the near-Earth asteroid Eros in 1898 allowed the parallax technique to be used on that asteroid. However, the results were not really any more accurate than the transit of Venus method which Halley had suggested many decades before.

In modern times Venus has been used in a totally different way to get a precise value for the Astronomical Unit. By bouncing radar pulses off the planet, astronomers have been able to measure its distance from the Earth, and then determine the distance from the Earth to the sun. This, the modern value for the Astronomical Unit, is 149, 597, 870.691 kilometres ± 30 metres.

To sum up, in the eighteenth and nineteenth centuries, transits of Venus were fundamentally important in scientific and technological development, prompting countries to compete and collaborate. Expeditions were sent to the ends of the Earth, with manifold geographical as well as astronomical results. Many of the expeditioners did not live to return to their mother countries, while others had adventures they would not forget in a hurry.

The primary purpose of Cook's mission – to determine the Astronomical Unit with great accuracy – failed in the short term. But as a result of his secondary instructions he mapped and claimed New Zealand and eastern Australia for the British crown. If he had not, perhaps these countries would today be French-speaking. It is an irony that Tahiti, where Cook carried out his historic observation of the 1769 transit, became a French colony. Perhaps, though, France deserved it after the hardships suffered by the expeditions of Chappe and Le Gentil?

***Duncan Steel,*** *a graduate of the University of London, was awarded his Ph.D. in 1985 by the University of Canterbury. Currently principal research scientist with the Ball Solutions Group, the Australian subsidiary of Ball Aerospace and Technologies Corporation, he is also a visiting researcher at the Australian Centre for Astrobiology at Macquarie University, and vice-president of the Spaceguard Foundation, which has its headquarters at the European Space Research Institute near Rome. He has worked on space missions for both NASA and ESA (European Space Agency), and is an expert on radar observations of meteors, optical tracking of asteroids and comets, orbital mechanics, the cosmic impact hazard, and the mathematics of calendars. He has written four books, over 130 scientific research papers, and hundreds of articles for magazines and newspapers, and appeared on many space-related television and radio documentaries. Minor planet 4713 Steel is named for him, as is a lunar-roving robot in Arthur C. Clarke's novel* The Hammer of God.

# Voyaging with Cook

*—Anne Salmond*

LET'S CAST OUR MINDS BACK to 1769, when Captain James Cook first arrived in the Pacific to observe this rare celestial phenomenon, the transit of Venus. After the victories of the Seven Years War, the Royal Society was determined Britain should show its supremacy by leading the international observations of the transit. In the eighteenth century these observations were of cosmological significance, intended to measure the distance of the sun from the Earth and thus establish the size of the solar system. To maximise their accuracy, it was decided that astronomers should be sent to widely spaced locations around the globe, and that one of these should be in the centre of the Pacific Ocean.

The Earl of Morton, then president of the Royal Society, made approaches to the King and the Admiralty, who agreed to fund this scientific mission. The Society chose James Cook, a master in the Royal Navy with a gift for astronomy and surveying, and appointed him as an observer. The Admiralty gave Cook a lieutenant's commission and a Whitby collier, the *Endeavour*, with orders to sail halfway around the world to observe the transit, and secret instructions to try and find Terra Australis Incognita, the Great Unknown Southern Continent, while he was at it. This fabled continent was a great prize, eagerly sought by European monarchs. It was thought that it must exist to counterbalance the great land masses of the north, or the world would topple off its axis, and that it would be rich in gold and silver, since these had been found in similar latitudes in South America.

Shortly after Cook's appointment, another British ship, the *Dolphin,* after circumnavigating the world, anchored in the Downs with news of a discovery that decided the *Endeavour*'s destination. Although Captain Wallis's men had been sworn to secrecy, the news soon got out that they had found a marvellous island in the Pacific with high mountains, bright rivers and waterfalls, coral reefs fringed with coconut palms, and beautiful, amorous women. The Royal Society decided that Tahiti would be a perfect location for the observations. They appointed Charles Green, an astronomer, as their main observer, while Joseph Banks, an ebullient, wealthy young sprig of the English élite with a passion for botany, offered to pay for the rest of the scientific party if he could go with them. The Royal Society accepted his offer with alacrity, and Banks joined the expedition, accompanied by Dr Solander, a favorite student of the great Swedish naturalist Linnaeus, an entourage of artists and draughtsmen, two black servants and a greyhound.

The party was lavishly equipped, for as one of Banks's friends reported to Linnaeus:

*No people ever went to sea better fitted out for the purpose of Natural History. They have got a fine library of Natural History; they have all sorts of machines for catching and preserving insects; all kinds of nets, trawls, drags and hooks for coral fishing. In short, Solander assured me that me that this expedition would cost Mr Banks 10,000 pounds.*

In addition, Cook was given by the Earl of Morton a detailed set of 'Hints', which suggested how he should conduct his observations and handle any 'Natives' he might meet. In particular,

Morton urged that the sailors not shoot any local people, for 'sheding the[ir] blood is a crime of the highest nature:– They are human creatures, the work of the same omnipotent Author, equally under his care with the most polished European. They are the natural, and in the strictest sense, the legal possessors of the several Regions they inhabit.'

The earl also suggested how the expedition's members might identify the Unknown Continent, study its Animal, Vegetable and Mineral Systems, and describe the Natives, beginning with their Persons, Features, Complection, Dress, Habitations, Food and Weapons.

The *Endeavour* was an Enlightenment voyage, the first of the great scientific expeditions into the Pacific. It was a travelling learned society, with its members mentally and morally prepared to collect and process the evidence of their discoveries: pressed plants; preserved birds, fish and animals; sketches and textual descriptions of these things and the people and places they encountered; artifacts and vocabularies, as well as charts, coastal profiles, logs and journals. And if people resisted their landings, they should not use their guns against them. Rather, the power of the British should be demonstrated by shooting a bird or animal, and sign language should be used to indicate what the sailors wanted – by holding up a jug, turning it upside down, showing them it was empty, then lifting it to their lips, for instance. Soft music might be played, and trinkets (especially looking-glasses) laid on the shore.

The *Endeavour* sailed to Rio, and then through the Straits of Magellan, heading for Tahiti.

Tahiti is a jewel of a tropical island – jagged volcanic mountains of black rock rising out of a blue-green sea, waterfalls tumbling through forests of flowering and fruiting trees, bright birds flying across the clearings. Like the sailors, the Tahitians were maritime explorers, although they saw the Pacific as a vast watery plain, joined around the horizon by the layered spheres of the sky, encircling its clusters of known islands. It was also a marae, a sacred place where people went to cleanse themselves in times of spiritual trouble.

According to Tahitian accounts, at the beginning of the world a generative force had produced first darkness, and then space, shooting stars and the moon, the sun and comets. As the star ancestors emerged one by one they sailed in canoes across the sky, and on their voyages of exploration new stars were created. The various entities of the world were generated, and eventually a star god created 'the kings of the chiefs of the earth . . . and the chiefs in the skies', each with their own star, whose boundaries were marked by a marae, a great stone temple. Thus, when Tahitian navigators sailed from a marae at the edge of the land, they were retracing the sky voyages of their star ancestors.

Tahitian histories say that in the mid eighteenth century a marae on the island of Ra'iatea – Taputapuatea – was dedicated to the war god 'Oro. From this temple, his followers carried the worship of 'Oro from island to island. They were led by the arioi, an exclusive society of priests, warriors, orators, famed lovers and voyagers. The war god demanded human sacrifices; his marae were dark, awesome places, regarded with dread and terror.

As the cult was taken by canoe across the archipelago, south to the Cook and Austral Islands, north to the Marquesas and east to the Tuamotu islands, Taputapuatea became the centre of a far-flung voyaging network. When the arioi travelled to the

A sketch, by Sydney Parkinson, believed to be of the great Taputapuatea marai at Opea, on the south-east coast of Ra'iatea, described in Joseph Banks's journal on 20 July 1769. A pig and two fish are placed on an elevated platform, or fata-rau. BRITISH LIBRARY

ceremonies at Taputa, their fleets of large carved canoes carried images of the gods, and pairs of dead men and sea creatures (including sharks and turtles) on their prows as sacrifices for 'Oro. As the priests landed at the marae, drums and conch trumpets sounded, and the bodies of the sacrificial victims were hung in trees by ropes strung through their heads, or laid as rollers under the keels of the sacred canoes as they were dragged up the beaches.

Although the 'Oro cult spread rapidly, some islands held fast to their old gods, and in about 1760, when warriors from Bora Bora attacked Ra'iatea, they chopped down one of the great trees that sheltered the sacred temple. Distraught at the desecration, a priest named Vaita entered a trance and announced that a new kind of people were coming to the islands:

*The glorious offspring of Te Tumu will come and see this forest at Taputapuatea. Their body is different, our body is different. We are one species only from Te Tumu. And this land will be taken by them. The old rules will be destroyed. And sacred birds of the land and the sea will also arrive here, will come and lament over that which this lopped tree has to teach. They are coming up on a canoe without an outrigger.*

News of this prophecy flew around the islands, and on 18 June 1767, when the first European ship, the *Dolphin* – a 'canoe without an outrigger' – arrived at Tahiti, Vaita's prophecy seemed to be vindicated. At this moment, Tahiti was 'discovered' and the Tahitians entered European history.

At the same time, however, the Europeans were entering Tahitian history, tangling these histories together. Wallis was searching for Terra Australis Incognita, hoping to inscribe its coastlines on the maps of the world, while the Tahitians thought the *Dolphin* had burst through the sky from another world or dimension. At first, they were not certain whether the strangers were people like themselves or supernatural creatures. More than a hundred canoes came out to look at the weird vessel. A priest made a long speech, then threw a plantain branch into the ocean. After this more canoes came out, and more speeches were made. The *Dolphin*'s crew did not understand a word that was said, but held up cloth, knives, beads and ribbons, and grunted like pigs and crowed like cocks to indicate what they wanted.

In the days that followed, the Tahitians decided to challenge the strangers. Warriors in a fleet of canoes attacked the *Dolphin*,

Confusion reigned in June 1767, when the first European ship, the *Dolphin,* under the command of Captain Samuel Wallis, sailed into Tahiti. After the islanders surrounded the boat in canoes, Wallis, fearing an attack, fired the ship's guns and killed many of them, an event depicted in this painting, believed to date from 1784. MICHAEL ANGELO ROOKER, 1743–1801, A-111-039, ALEXANDER TURNBULL LIBRARY

hurling stones with slingshots. The sailors ran to their guns and opened fire, killing many of them. Wallis sent ashore an armed party, who 'took possession' of the island, hoisting a red pennant (in Tahiti, the colour red was a sign of 'Oro's presence). Later that evening, two old priests ceremoniously lowered the pennant, and took it away. The next morning, a huge crowd paraded with the red banner flying on a high pole and surrounded the ship's boats at the watering-place. Wallis, fearing that his men were about to be attacked, fired the ship's guns again, killing many more people.

George Robertson, the ship's sailing master, wrote in his journal:

> *How terrible must they be shocked, to see their nearest and dearest of friends dead, and torn to pieces in such a matter as I am certain they never beheld before. To attempt to say what these poor ignorant creatures thought of us, would be taking more upon me than I am able to perform.*

From this time on, the power of Wallis and his men was no longer seriously challenged. The local people came to make peace, and offered their women to the sailors. Arioi warriors were sexually voracious, and the Europeans were probably being treated in this fashion. Purea, the 'queen' of Tahiti and a leading arioi, appeared with her 'right-hand man' Tupa'ia, intent on forging a close relationship with the Europeans.

Tupa'ia was a leading arioi, a high priest navigator from Ra'iatea who had brought the worship of 'Oro to Tahiti, and after this first meeting with the British he and Purea spent a good deal of time with Captain Wallis's men. They went on board the *Dolphin* on a number of occasions, and Purea invited Wallis to her home. In a ceremony at the great arioi house, to seal the alliance she gave him gifts and put a bunch of sacred red feathers on his hat. After Wallis and his men left the island, the red pennant from the *Dolphin* was made into a feathered girdle for Purea's family. With this girdle, along with others representing high titles, she intended to install her son as paramount chief of the island.

Some months later, when two French ships commanded by Bougainville arrived on the east coast of the island, Purea and her husband Amo were busily building a huge marae at Papara on the west coast, to the consternation of their rivals. When Purea and Amo announced their son's installation, the tribes to the south and north-west were angered, and in December 1768 they attacked Papara. Devastating raids by the southern warriors, which led to great loss of life, were followed by an onslaught from the north and the west. The warriors from the west seized the 'Oro image from the new marae and the feather girdle incorporating Wallis's ensign. Tupa'ia, with Purea and all of her family, fled to the mountains, while their enemies laid waste to the district. When the warriors finally returned to their homes, Purea's family and Tupa'ia were allowed to return to Papara, but their power had been greatly diminished.

THIS WAS THE STATE OF AFFAIRS when, in April 1769, the *Endeavour* sailed into Matavai Bay. Cook was in command, accompanied by Joseph Banks and his Royal Society entourage of artists and scientists. When Cook and his companions went ashore, hundreds of the 'meaner sort' of Tahitians stood on the beach, holding up plantain branches. An envoy approached the British on hands and knees, presenting his branch to Cook, and Lieutenant Gore, a *Dolphin* veteran, suggested that each of their party pick up a plantain branch before walking through the trees to find the great house where Purea had entertained Captain Wallis. When they arrived at the site, however, Gore was stunned to find that the fine house, the stone temple with its carvings, the pigs and chickens, and the lush gardens had all vanished, as had the large canoes that had been in the bay. During the recent battles the area had been devastated.

Early the next morning, two distinguished elders came to escort Cook and Banks ashore. They dressed them in their own bark-cloth garments, imbued with their personal mana, and after landing on the beach led them to Tutaha, the leader of the Pare-Arue district who had defeated Purea's people just six months earlier. Tutaha presented Cook and Banks with two lengths of perfumed bark-cloth, and a cock and a hen. When Banks gave the chief his lacy silk neck-cloth in return, Tutaha seemed delighted. Fearing that Purea would reestablish her friendship with the British, he had hastened to establish his own relationship with their leaders.

After this ceremony, some of Tutaha's women offered sexual hospitality to their visitors with unmistakable gestures but, much to Banks's chagrin, Cook refused. Soon afterwards they met another leading chief, Te Pau, who forged a close relationship with Banks in the days that followed. No doubt the local chiefs were anxious to discover Cook's intentions. Captain Wallis had been Purea's ceremonial friend, and they must have expected the British to punish them for their attack on her people. The 'queen' did not appear, however, and in her absence they met other local leaders.

Relationships were excellent until one day, while Cook and Banks were away looking for supplies, a man snatched a sentry's musket at the encampment and was shot dead, while others were wounded. All the Tahitians fled, and when Banks heard what had happened he was furious, exclaiming, 'If we quarrelled with those Indians, we should not agree with angels.' He crossed the river and met with the *Dolphin*'s old man, who persuaded others to return with them, bearing plantain branches.

After this peacemaking the sailors began to build a fort, and Banks, his companions and many of the sailors began to spend most of their time ashore. As Molyneux, one of the *Dolphin* veterans on board the *Endeavour*, remarked:

> *the Gentlemen off duty take a great deal of Pains to learn the Language of the Country & have made some progress therein. Individuals form Freindships with Individuals & every man has his Tayo (or Friend); this might be productive of good consequences but the women begin to have a share in our Freindship which is by no means Platonick.*

Finally, on 28 April, a large fleet of canoes arrived at Matavai Bay, bringing Purea, Amo and their high priest Tupa'ia. On meeting Molyneux, Purea confided that during his absence she had been attacked and dispossessed of all her lands. The 'queen' went out to the *Endeavour*, where Cook invited her into the great cabin and presented her with a doll, whimsically saying it was an image of his wife. Purea was delighted, and held it up as she returned ashore. Tutaha, thinking this was a particular sign of favour, was affronted until he was also presented with a doll, which no doubt seemed like an ancestral image or ti'i. He must have been worried that the British, having met Purea and heard her story, would reestablish their alliance.

Cook tried to stay neutral in these struggles for power, but Banks was much less discreet and over the following days he began a passionate affair with one of Purea's women. Relationships were increasingly intimate, but still volatile. When the astronomical quadrant, which was kept under close guard in the fort, was stolen, Banks and Cook set off in pursuit of the thief, and

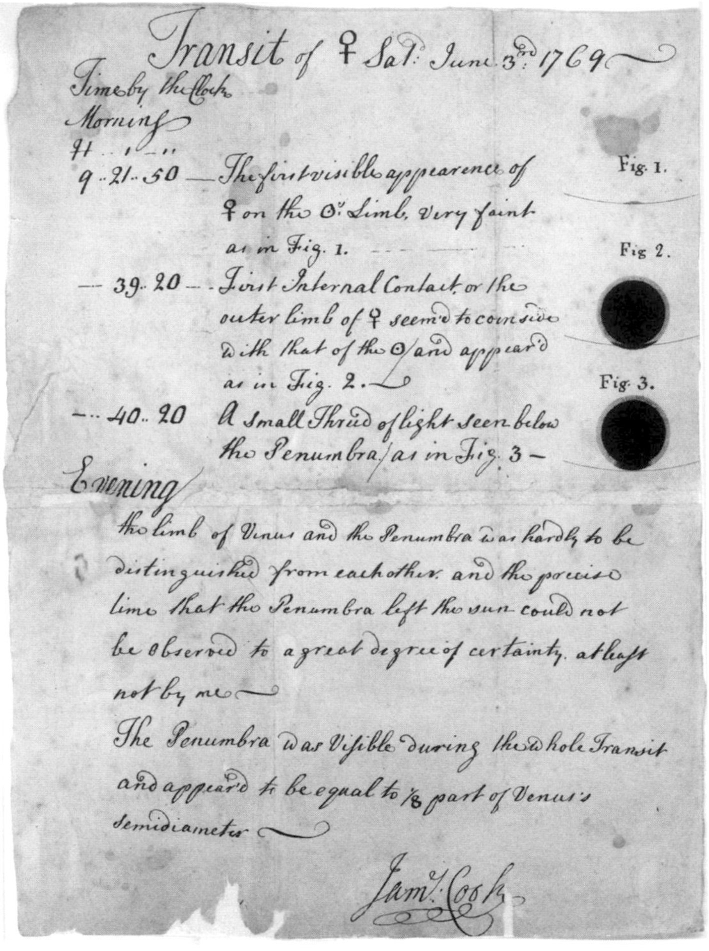

Transit of ♀ Sat: June 3rd 1769

Time by the Clock

Morning

H ' ''

9..21..50 — The first visible appearence of ♀ on the ☉'s Limb, very faint as in Fig. 1.

— 39..20 — First Internal Contact, or the outer limb of ♀ seem'd to coincide with that of the ☉ and appear'd as in Fig. 2.

— ..40..20 A Small Thread of light seen below the Penumbra as in Fig. 3 —

Fig. 1.

Fig. 2.

Fig. 3.

Evening

the limb of Venus and the Penumbra was hardly to be distinguished from each other. and the precise time that the Penumbra left the sun could not be observed to a great degree of certainty. at least not by me —

The Penumbra was Visible during the whole Transit and appear'd to be equal to 1/8 part of Venus's semidiameter —

Jam:s Cook

This page from Captain James Cook's diary describes his observation of the transit of Venus from Matavia Bay, Tahiti on 3 June 1769. Cook had developed an interest in astronomy, and three years ealier, in 1766, while surveying the Newfoundland coast, had witnessed an eclipse of the sun and sent a report on it to the Royal Society.

MS-PAPERS-0200-24-1, ALEXANDER TURNBULL LIBRARY

in their absence Tutaha, who had tried to flee the bay in his canoe, was captured and roughly handled. When they returned with the quadrant – which, much to their dismay, had been dismantled – Tutaha was released, and presented Cook with two pigs, seething with rage about the way he had been treated.

From this time on, Tupa'ia spent much of his time with the British, especially Banks and his party. He must have been gratified with the way his enemy had been humiliated, and certainly he was fascinated by Banks's retinue, with their scientific and artistic equipment. Banks was wealthy and well-born, with elegant clothes and an amorous disposition – just like an arioi – while Tupa'ia was one of the most intelligent and knowledgeable men in the archipelago. He was not only a high priest, but a navigator who had travelled widely though the islands. As their mutual command of each others' languages improved, he tried to teach Banks and his companions about Tahitian navigation, the location of islands in the surrounding seas, and Tahitian beliefs and customs. He and the other Tahitians also began to learn the sailors' names, although they found the consonants difficult to pronounce – Cook became Tute, Banks Topane, Solander Tolano, and Parkinson Tatini.

Having befriended Joseph Banks, Tupa'ia acted as guardian and guide to Banks's companions, instructing them in local etiquette and joining in their rituals, including Divine Service. When Banks slept in Purea's canoe one night, and his white jacket and waistcoat (with silver frogs) were stolen, Tupa'ia stood guard over his musket. Some time during these few weeks, Tupa'ia learned to paint in the English style, either from Sydney Parkinson or Hermann Spöring, Banks's artists (although Spöring was more properly a draughtsman). There are a number of naïve sketches from the voyage, which have previously been attributed to Joseph Banks. A letter from Banks recently came to light, however, which states that Tupa'ia drew one of them:

> *Tupia the Indian who came with me from Otaheite Learned to draw in a way not Quite unintelligible. The genius for Caricature which all wild people Possess Led him to Caricature me & he drew me with a nail in my hand delivering it to an Indian who sold me a Lobster.*

Some male arioi were skilled painters on bark cloth, and Tupa'ia may have been one of them. It is suggestive that his images use red, brown and black, the predominant colours of bark-cloth painting. The motifs, however, are more likely to have come from Tahitian tattooing, where naturalist images of plants and people were common. In his turn, Sydney Parkinson learned the names of the Tahitian dyes and the plants from which they were made, and both he and Joseph Banks acquired tattoos of their own.

After learning to sketch in the European style, Tupa'ia worked with Banks and Cook on charts of Tahiti and Ra'iatea, dictating place-names to be written along the coastlines. When the *Endeavour* sailed from Matavai Bay after three months in Tahiti, Tupa'ia sailed with them, acting as the ship's navigator during their journey through the Society Islands. He chanted to raise the winds, took the ship safely through channels in the reefs, led them through the proper rituals upon landing on each island, and finally guided them to the sacred marae of Taputapuatea, the headquarters of the arioi. As an experienced navigator, Tupa'ia had a detailed knowledge of the sea and sailing conditions, which allowed him to move with relative ease between the European and Tahitian maritime traditions, communicating effectively with the sailors. Nevertheless, he remained a high priest of 'Oro: from the time they left Tahiti, Tupa'ia was taking the *Endeavour* on an arioi voyage.

After leaving the Society Islands, the *Endeavour* sailed south for New Zealand. During this passage Tupa'ia worked with Cook to produce a chart of the islands around Tahiti, indicating their locations, whether or not they were inhabited, and describing their topography and resources. In addition to this remarkable chart, he dictated a list of 130 islands with which he was familiar, including ones in the Tuamotus, the Marquesas, the southern Cooks and the Samoan archipelago, and Pounamu and Teatea, the North and South Islands of New Zealand, although he had visited only 12 of these, including eight in the Society Islands, two in the Australs and two in the Tongan archipelago.

As they sailed south, Tupa'ia must often have spoken with 'Oro in his dreams, and watched the night voyages of his star ancestors. During the *Endeavour*'s six-month circumnavigation of New Zealand, his presence on the ship greatly impressed local Maori. He was a high priest from Ra'iatea, a fabled homeland of their ancestors, who could understand their language. He often spoke with their priests, and these must have been marvellous conversations. At Uawa on the East Coast, Tupa'ia slept on shore, left a painting in a cave, and visited a carved house on an offshore island. A carving from this house, an ancestral portrait, was probably presented to Tupa'ia. When the *Endeavour* returned to England, the carving was painted in London and then vanished, reappearing a few years ago in

Tuebingen Castle, where it was identified from its eighteenth-century portrait.

When Cook returned to New Zealand on the second voyage, and the canoes came out to the ship, their crews cried out for Tupa'ia, and wept bitterly when they heard he had died in Batavia. It seems Maori thought the *Endeavour* was Tupa'ia's ship, and they mourned and remembered him for generations.

During many of the early European voyages in the Pacific, it seems to me, some of the deepest cross-cultural encounters were not between Polynesians and Europeans but between the inhabitants of different Polynesian islands. These were, after all, Pacific seas they were traversing, and Pacific places they were visiting.

In the past, the power of people such as Tupa'ia to shape the trajectory of these voyages has been greatly underestimated. For over a decade in the Pacific, the 'travelling cultures' of Cook's ships were radically altered. Polynesians joined the ships for long periods, and travelled to many places. They learned to speak English, to eat European foods and wear European clothes, and had some very exotic experiences, including those with other Pacific Islanders. Some Europeans acquired tattoos, and had Polynesian friends and lovers. They learned to like Pacific foods, and became bilingual in English and Maori (or Maohi). In their successive encounters, European voyagers and Polynesians alike were transformed. Tupa'ia and other Polynesians became entangled with Europe. Cook and Banks and other Europeans became entangled with the mana and tapu of the Pacific. I believe Cook and many of his men changed in the process, and that this led to his death – although that, as they say, is another story.

The story of the transit of Venus and its impact on Pacific history is, thus, a cross-cultural one. It is fascinating that it was the knowledgeable experts on both sides – the scientists from the Royal Society, particularly Joseph Banks, on the one hand, and the arioi high priest, Tupa'ia, on the other – who moved most readily across linguistic and cultural boundaries, and engaged with each other. It must have been sheer intellectual curiosity which impelled them, and a shared fascination with new knowledge.

In turn, their entanglements helped determine what happened during the voyages. In accounts of maritime explorations, therefore, and the cross-cultural encounters they involved, the historicity of indigenous protagonists cannot be ignored, just because they were not European. If we do so, our histories become hollow, mere imperial echoes. For, as the Tahitian prophet Vaita proclaimed, 'We are one species from Te Tumu,' and good historical scholarship is grounded on that common humanity.

Or as my mentor, the Maori tribal expert Eruera Stirling, used to chant about his great ancestral homeland, Hawaiki:

Whakarongo! Whakarongo! Whakarongo!
    *Listen! Listen! Listen!*
Ki te tangi a te manu e karanga nei
    *To the cry of the bird calling*
Tui, tui, tuituiaa!
    *Bind, join, be one!*
Tuia i runga, tuia i raro,
    *Bind above, bind below*
Tuia i roto, tuia i waho,
    *Bind within, bind without*

Tuia i te here tangata
*Tie the knot of humankind*
Ka rongo te poo, ka rongo te poo
*The night hears, the night hears*
Tuia i te kaawai tangata i heke mai
*Bind the lines of people coming down*
I Hawaiki nui, I Hawaiki roa,
*From great Hawaiki, from long Hawaiki*
I Hawaiki paamamao
*From Hawaiki far away*
I hono ki te wairua, ki te whai ao
*Bind to the spirit, to the day light*
Ki te Ao Maarama!
*To the World of Light!*

***Anne Salmond** is a Distinguished Professor in Maori Studies and Anthropology at the University of Auckland, and the author of six award-winning books on Maori life and early contacts between Europeans and islanders in Polynesia. She is a Fellow of the Royal Society of New Zealand, a Dame Commander of the British Empire, chair of the New Zealand Historic Places Trust, and recipient of the Prime Minister's Award for Literary Achievement. Her most recent work,* The Trial of the Cannibal Dog: Captain Cook in the South Seas, *won a Montana New Zealand Book Award in 2002. She is currently researching a new book,* Aphrodite's Island: The European Discovery of Tahiti.

**Sources of quotations**

Page 80 Beaglehole, J.C., ed., 1955, *The Journals of Captain James Cook on His Voyages of Discovery, Vol I: The Voyage of the* Endeavour *1768–1771* (Cambridge University Press for the Hakluyt Society), cxxxvi.

Page 81 Ibid, 514–515.

Page 82 Henry, Teuira, 1907, 'Tahitian Astronomy (Recited in 1818 at Porapora, by Rua-nui [Great-Pit], a clever old woman) Birth of the Heavenly Bodies,' *Journal of the Polynesian Society* 16:101–104.

Page 84 Driessen, Hank, 1982, 'Outriggerless Canoes and Glorious Beings', *The Journal of Pacific History* XVII:8–9.

Page 86 Robertson, George, ed. Oliver Warner, 1955, *An Account of the Discovery of Tahiti, From the Journal of George Robertson, Master of HMS* Dolphin (London, Folio Society), 43.

Page 88 Parkinson, Sydney, 1773, *A Journal of a Voyage to the South Seas in His Majesty's Ship* Endeavour (London, for Stanfield Parkinson), 15.

Page 89 Molyneux, Robert, *Log*, Adm 55/39, Public Record Office, London, f.58.

Page 92 Carter, Harold, pers. comm. 1997.

# Travels in Time and Space

*—Paul Callaghan*

ONE OF THE OBSERVERS of the 1882 transit of Venus, William Harkness of the United States Naval Observatory, wrote:

> *There will be no other till the twenty-first century of our era has dawned upon the Earth, and the June flowers are blooming in 2004. When the last transit occurred, the intellectual world was awakening from the slumber of ages, and that wondrous scientific activity which has led to our present advanced knowledge was only just beginning. What will be the state of science when the next transit arrives, God only knows. Not even our children's children will live to take part in the astronomy of that day.*

Those of us living in this century do, indeed, have the extraordinary good fortune to be able to witness not just the recent transit of Venus of 8 June 2004, but its 'twin' transit, which will take place on 5 June 2012, at 10.15 in the morning. A transit of Venus is, in effect, a metaphor for the great voyage of science, a voyage which began as a glimmer in the human mind about 2500 years ago, forged its underlying navigational tools in ancient Greece, and was launched in full vitality during the European Renaissance which began in the fifteenth century. It is a voyage that carries us on through the twenty-first century, and is, in a real sense, *the* voyage of the future, one that will take the human race to places and through experiences that will rival those of Odysseus.

At the 1882 waypoint, when electricity was a new curiosity and air travel still a fantastic notion, William Harkness believed 'that wondrous scientific activity which has led to our present advanced knowledge is only just beginning'. And of the prior transit in 1769, he wrote: 'The intellectual world was awakening from the slumber of ages.'

The slumber of the ages indeed. In the 100,000 years that mark humanity as we think we know it, *Homo sapiens* had discovered agriculture, built communities and great cities, constructed ships, bridges and cathedrals, and fought with great armies. They had developed complex technologies based on bronze and then iron, and, in China, explosives. There was technology. There was empirical experience. But there was no science. This was a pre-Renaissance world, where things were explained, where possible, by common sense. When that broke down, mysticism reigned and the role of the gods was invoked.

What we now call 'science' is western science. All science as we now know it had its beginnings in Greece. In the words of Albert Einstein:

> *The development of western science has been based on two great achievements: the invention of the formal logical system by the Greek philosophers, typified by Euclidean geometry; and the discovery during the Renaissance of the possibility of finding out causal relationships by systematic experiment. One need not be surprised that other civilisations, such as the Chinese, did not make these steps. The astonishing thing is that these discoveries were made at all.*

So what is science? I like the view of biologist and writer Lewis Wolpert the best: Science is a way of looking at the world that tries to explain natural phenomena in terms of underlying causes, in a way that is self-consistent and corresponds with reality. Archimedes' law is independent of culture or religion. It is neither good nor bad. It is simply true. Social relativism is anathema to science.

But in practice there is no agreed definition of science, and we scientists get along perfectly well without one. Science takes place in a chaotic manner, nothing at all like the neat objective style in which ideas are subsequently presented in scientific papers or textbooks. It is a mixture of creative thought and detailed, often tedious, work. The best scientists are sloppy enough to allow for unexpected outcomes, but organised enough that they can find out what happened. Science is revolutionary. It builds a solid core of established knowledge, but at the frontiers it is in a state of restless upheaval.

There is something extraordinary about science that explains why its very discovery was, in Einstein's words, 'astonishing'. It is an aspect that is central to those of us who both practise science and teach it. And it is this: science is a system of discovering knowledge that defies common sense. It is, after all, common sense that the sun goes around the Earth. It is common sense to say that objects need forces in order to move. It is common sense to say that continents don't move and that animal species are immutable. It is common sense to say that if I throw a coin four times it is more likely I will get head-tails-heads-tails than heads-heads-heads-heads. Yet all these common-sense ideas are wrong.

Evolution confounds common sense. So do Newton's laws, vaccination by cowpox to prevent smallpox, and our biological inheritance being contained in the code of four base pairs on a long molecular double helix. And, while we are thinking about molecules, the fact there are more molecules in a glass of water than glasses of water in all the oceans confounds common sense. There was a time, too, when the idea that the Earth and Venus orbited the sun in concentric but slightly inclined orbits confounded common sense.

But all these ideas form the basis of a deep understanding that allows us to overcome mysticism, unravel the underlying patterns of nature, and take a measure of control over our lives, our health, our ability to feed ourselves, to communicate, to travel, and to free ourselves from drudgery.

That utility, however, is not the reason for the voyage of science. We have undertaken the journey because one of our deepest human yearnings is to know, understand and reach out to the universe around us, and to place ourselves in its context. I want to look to where that journey might take us.

Now I have to say that such crystal-ball gazing is fraught with dangers. We need only look to past experience to see the pitfalls. It was, for example, a commonly held view at the end of the nineteenth century and the beginning of the twentieth that science had a fairly complete picture of nature's laws. How wrong. The new physics of quantum mechanics would usher in the age of chemistry, of electronics, of nuclear power, and of the myriad new techniques, such as X-ray diffraction and magnetic resonance, which would enable scientists in the second half of the twentieth century to unravel the structure of proteins and DNA, and understand the molecular basis of life itself.

Observations of transits of Venus over three centuries helped establish the distance of the sun from the Earth, but it was New Zealand physicist Ernest Rutherford who cracked the mystery of the Earth's age, applying knowledge of radioactive decay to common elements.
PHOTOGRAPH BY UMBO, SIR E MARSDEN PAPERS, PACOLL-0091-1-003, ALEXANDER TURNBULL LIBRARY

But, in case you think that there was something orderly and planned about all this, let me hasten to assure you that, as always in science, chaos and unpredictability ruled supreme. Take the case of Ernest Rutherford. As a physicist, Rutherford discovered the structure of the atom and the nature of nuclear radioactive decay. But in effect he laid the foundations for modern chemistry: with his nuclear model it became possible to explain Mendeleev's periodic table of the elements.

Rutherford also, incidentally, laid to rest one of the greatest controversies in Earth science and biology, the true age of the Earth. He saw that by analysing the isotope ratios of common elements and applying the rules of radioactive decay, he could calculate that age. He came up with an answer that was orders of magnitude greater than anyone had believed. Using his method the age is 4500 million years, a value that has stood the test of every subsequent estimate.

By studying the nucleus of the atom, Ernest Rutherford had profoundly advanced chemistry and geology. But he had not the slightest idea of the nuclear uses of his discoveries, famously dismissing any possibility of nuclear energy extraction as 'moonshine'. Nor did he realise what he would contribute to astronomy. It would be another New Zealander, Beatrice Hill Tinsley, who would use nuclear physics to understand much about how stars are born, how they shine and how they die.

Edward Purcell, the American physicist who co-discovered nuclear magnetic resonance in 1945, thought it might be useful for measuring magnetic fields, whereas in fact it revolutionised synthetic chemistry, provided a way to find protein structures, and gave us new medical scanners. In fact, nuclear magnetic resonance has generated quite bizarre technology applications, including explosives detection and prototype quantum computers.

And speaking of computers, the brilliant John von Neumann, who conceived of the idea, had not the slightest inkling of how ubiquitous computers would become, or that most of us would spend a significant fraction of our lives interacting with them.

In short, the discoverers of new science are generally the least likely to see its potential. Did New Zealand's Maurice Wilkins, ever in his wildest imagination, dream that his DNA work would lead to a whole new way of connecting the human family? Could he have conceived that another New Zealander,

Allan Wilson, working at Berkeley, would find a way to use mitochondrial DNA to discover the links between us all, and the pathway to a common maternal ancestor somewhere in Africa around 150,000 years ago, in effect giving birth to new fields of science, genetic history and genetic anthropology?

And if these scientists could not foresee the science future, could the policy leaders do a better job? In 1937, the United States National Academy of Sciences organised a study aimed at predicting breakthroughs in the next 50 years. They had some interesting successes, such as predicting new uses for synthetic polymers. But here's what they overlooked:

- antibiotics
- nuclear energy
- space exploration
- DNA and protein structure discoveries
- computers.

That should be enough to make anyone tremble at making predictions, but it didn't stop *Scientific American* writer John Horgan, who produced as his millennium contribution a book called *The End of Science*, in which he argued, in *fin de siècle* style, that all the great questions of science had been answered and all that remained in the twenty-first century was to work out the applications

In my view, Horgan was profoundly wrong. As Rutherford looked further into the atom he discovered the nucleus. As later physicists looked further they discovered the substructure of protons and neutrons, and then *their* substructure of quarks, only to find mystery in the nature of the symmetries that overarch the forces of nature. We do not understand the true nature of those forces, nor why nature gives us an endless zoo of elementary particles that seem completely unnecessary to any human-centred life. In short, if God had made the universe for us, why would he have made the tauon and the tauon neutrino? Are they meant for some other universe? We continue to face questions that go to the heart of basic physics.

And what as we look outwards from Earth? As Galileo and his Renaissance contemporaries looked into the heavens they discovered the details of our solar system, and the moons around the various planets. Later, each improvement in telescopes allowed us to see further through the stars of our Milky Way galaxy, and as telescopes grew more powerful still, the strange spiral objects in the sky were revealed as more distant galaxies.

When Edmond Hubble found that distant galaxies were receding, he realised the universe was expanding. That, by extrapolation, led others to suggest a 'big bang' origin. But the big bang itself raises immense questions about the nature of space and time, and in particular the first instant of the universe, when all the forces were unified. Hubble's work gave impetus to the new field of cosmology, an area of science where the physics of elementary particles and massive gravitational objects find a unity, general relativity meets quantum mechanics, and the black hole provides a scientific journey for the mind, a place we cannot approach except through our scientific and mathematical imagination.

It was to this place that a young New Zealand mathematician, Roy Kerr, ventured in his mind in the 1960s to solve a problem that had eluded everyone else: How could one describe in mathematical terms the physics that prevails when a black hole rotates? Kerr's contribution to understanding black holes stands alongside that of Einstein, who laid the foundations, and Stephen Hawking, who brought to black holes the methods of quantum mechanics and thermodynamics.

And now we are in the era of the orbiting Hubble telescope, with which we can see the birth of galaxies at the edge of the universe. New space probes have given us a glimpse of the faint dappling in the microwave background radiation (a texture that might have been appreciated for its poetic possibilities by Gerard Manley Hopkins), which raises questions about fluctuations that occurred at the instant of birth of the universe 15 billion years ago. Even more mysteriously, our observations now tell us of a great fraction of missing mass in what we call the dark matter, about whose meaning we can, at this point, only speculate.

And so the science of the cosmos beckons us with big questions about the nature of forces, time, matter and space.

AND WHAT OF BIOLOGY? We are starting to see how genetics drives the complex interweavings of evolution, speciation – the formation of new species – and phenotypes, the physical and behavioural characteristics of individuals and groups. Our understanding of disease is changing fast as we learn not only of the influence of genetic factors, but also of the role viruses play in diseases previously thought to be degenerative. The nature of ageing in humans, and precisely why and how it occurs, is now a matter of open debate.

As we use our new atomic and molecular tools we see the mechanisms which drive the replication processes of life, and we grasp for a new science at the interface between physics, chemistry and biology, a type of physics and associated mathematics that can handle the complexity – that can accommodate out-of-equilibrium dynamics, chaos, self-organisation and punctuated avalanche effects. We don't know that physics yet. It is one of the biggest challenges facing people such as me, who study complex soft matter.

And as we peer into a single human brain we find that it has more neurons than the billions of stars in the Milky Way. We see the glimmerings of how memory works, and we start to wonder how emotions are encoded and how creativity is part of human consciousness.

The end of science? I don't think so.

As we peel back each layer of understanding, science appears to draw us on in ever-expanding complexity, new measurement tools driven by new technologies driven by new science, in an unpredictable endless cycle.

Physicist Freeman Dyson has likened the journey of science to geographical exploration. In his 1989 book *Infinite in All Directions* he wrote:

> *Our vision can be dominated by the mountain tops. By the big picture, by the grand underlying principles. But in between the mountains lies the jungle. And the jungle is complex, vibrant, and infinitely diverse.*

And so I want to try to look ahead, to see in some trepidation where we might go. But first I want to ask: Why do we want to travel? Why are we on this journey? I started by saying that science is worth the exploration for its own sake, because it is there. But not everyone sees intrinsic value in science. D.H. Lawrence wrote:

> *The universe is dead for us, and how is it to come alive again? 'Knowledge' has killed the sun, making it a ball of gas with spots. 'Knowledge' has killed the moon ... it is a dead little earth fretted with extinct craters as with smallpox ... The world of reason and science ... this is the dry and sterile world the abstracted mind inhabits.*

This is to me a deeply pessimistic perspective. I prefer the optimism of the Nobel Prize-winning physicist Richard Feynman:

> *Poets say science takes away from the beauty of the stars – mere globs of gas atoms. Nothing is 'mere'. I too can see the stars on a desert night and feel them. But do I see less or more? The vastness of the heavens stretches my imagination – what is the pattern, the meaning, the why? It does not do harm to the mystery to know a little about it. Far more marvellous is the truth than any artists of the past imagined.*

When Beatrice Hill Tinsley was able to explain how nuclear-reaction cycles make the stars shine, she answered the question that had been in the mind of every human who, since time immemorial, has walked out on a clear night and gazed with wonder at the myriad stars in the heavens. Does her insight lessen that beauty, or enhance it?

Think of Maurice Wilkin's DNA: two metres in every cell, with 3.2 billion letters of coding; each of us with 10,000 trillion cells stretching 20 million kilometres.

Think of that DNA coding for our proteins: proteins made from 20 different amino acids, with each protein self-assembled

New Zealand astronomer Beatrice Hill Tinsley, who became professor of astronomy at Yale University, did groundbreaking work on the evolution of galaxies, and proved that the universe was not a closed system but was expanding.

into a special shape to allow it to function; a typical living cell having 20,000 different types of protein, each one with thousands of variants, and 100 million protein molecules, always in action, pumping ions across the cell membrane, managing the energy flows in the mitochondria, repairing damaged DNA, giving cells special markers so your immune system knows it belongs to you, providing the molecular motor of muscle cells.

Does that insight lessen or enhance our sense of wonder at the beauty of life?

Samuel Johnson said, 'Poetry is the art of uniting pleasure with truth.' And so I ask: Is science poetic? Is it poetic so that we should love it? I believe it is. Like Feynman, I feel science takes nothing from my sense of magical wonder at nature. Science only enhances that wonder.

What makes scientific poetry so different, though, is that we cannot write it with total freedom of the imagination. Let me quote from Feynman again:

> *Our kind of imagination is quite a difficult game. One has to have the imagination to think of something that has never been seen before, never been heard of before. At the same time the thoughts are restricted in a straitjacket, so to speak, limited by the conditions that come from our knowledge of the way nature really is. The problem of creating something which is new, but which is consistent with everything which has been seen before, is one of extreme difficulty.*

And so with extreme difficulty, and trying to be consistent with everything that has been seen before, let me attempt to gaze into the future to see where our poetic science voyage might take us. In doing so, I will try to imagine the technological consequences, without wishing to suggest that these changes will improve the lot of humanity.

Let me begin with my own subject, physics. I see the extension of quantum mechanics so we can build entangled quantum states over long distances, ushering in the possibility of quantum teleportation of information, and the age of the quantum computer. Such computers, should we achieve them, will be so astoundingly powerful they will challenge our whole notion of what a computer is.

I see the expansion of our ability to build structures at the atomic scale, making the era of nanotechnology a reality. This will allow computer speed and memory capability to grow exponentially. Nanotechnology will make possible a much greater range of unmanned space exploration, using new space probes that are tiny, multiple and light, further extending our search for extraterrestrial life to Jupiter and Saturn's moons, perhaps further.

It will make possible the fantastic journey of medical aids inside our body, targeting cancer cells with drugs and carrying out local tissue repair.

It will cause a revolution in information technology and personal communication electronics – the cybersphere internet. Wherever we are, we will remain completely digitally connected to the world, and to information about where we are, what we have done, what we will do, and whatever is possible to be known.

I see further great advances in astronomy, using new arrays of space telescopes, as well as a much larger Hubble beyond the moon. There will be new advances in astronomy based on gravitational waves, and much deeper insight regarding the nature of the universe and its mysterious dark matter, as well as its origin and end-point.

I see new developments in the physics of complex systems: physics that will allow us to understand how nature works in its apparently chaotic manner, with outbursts of activity separated by punctuated equilibria; physics of that highly non-linear type of behaviour known as self-organised criticality; physics of earthquakes and of turbulence; physics of the brain, and of the life process.

I see remarkable developments in software. Today's young adults, born in the mid 1980s, have in their lifetime seen the commencement and subsequent power of the internet, the demise of floppy disks, film and video, and their replacement by vastly improved digital technology. Yet in software, apart from the internet, they have seen little change. The programmes we use for word-processing, email and handling data have hardly changed since 1985, and nor has the graphical user interface – the screen, the windows, the menus, the mouse and all the other components. Despite vast improvements in hardware, software has grown only more cumbersome. Computers are now the only industrial products that are expected to fail. This software brittleness must end soon with a revolution in adaptive software, software that can adjust and learn.

Increasingly, too, we will come to think of genetic science as pure information technology. When we copy an antifreeze gene from an Antarctic fish and paste it into a tomato, we will see this as no more than writing software. Our genetic and proteomic database capacity will grow remarkably. And in genetic science we will see staggering applications as we learn to compute the function of the proteins expressed by genes.

Just as Moore's Law tells us that the power of computers increases exponentially for every dollar spent, Richard Dawkins has identified a 'son of Moore's Law' by which the numbers of base pairs of DNA that can be sequenced similarly rises per dollar spent. This law predicts that by 2050 the entire genome of any individual will be able to be sequenced for a few hundred dollars. A doctor may then be able to analyse your entire health profile, giving you a prescription which precisely fits your genome and, alarmingly, predicting with accuracy your eventual demise. Police may be able to take blood or semen from a crime scene and reconstruct in the computer the face of its owner at any selected age during his or her life.

The implications for understanding evolution will be astounding as we come to grips with the entire phylogenetic tree. We will be able to implement the dream of Nobel Prize-winning physiologist Sidney Brenner to reconstruct a detailed picture of the common ancestor of man and chimpanzee. Indeed, we may be able to reconstruct life itself from a knowledge of the software. *Jurassic Park* looks less absurd as each year passes.

But it is in medical science that some of the most remarkable forays of the scientific odyssey will occur. The power of prostheses – artificial body parts – will grow as, using the new tools of nanotechnology, we cross the silicon–human tissue bridge. We will move beyond cochlear implants to retinal implants, restoring sight to the blind. We will connect disconnected spinal nerves. We will use neural implants in our brain or spinal cortex to interface with the cybersphere around us. As biological augmentation takes over we will cross the barrier from brain to machine.

The brain will become the great frontier of science as we struggle to understand it, using new functional imaging techniques, new knowledge of brain biochemistry and its genetic base, and new knowledge of the way this vast neural network functions. At the same time, developments in drugs based on genomic insight will enable us to target specific brain function. We will, in effect, be able to manage memory. A frightening prospect is the recreational use of drugs taken in advance of pleasurable activity, to enhance memory, and the opposite, memory-suppressing drugs taken when unpleasant tasks are to be performed.

Macbeth pined for 'some sweet antidote to sorrow'. Yet as we contemplate such a prospect, we leave behind the world of humanity as Shakespeare knew it. And what do we become? To what extent will we remain human in circumstances where we start to understand the brain, how memory works, and that emotion, kindness, empathy, ambiguity, moral judgement, creativity and love are all simply functions of our nerve cells? To what extent will we remain human as we learn how to modify these essential features of our very selves? As we cross the barrier of tissue and prosthesis, as we connect our brains directly to the computer and modify our thought processes by drugs, do we lose life's specialness, what we have in the past called 'a reverence for life'?

Until now, these domains of utmost complexity – the biosphere, the human spirit, the life force – have been the province of indigenous knowledge and religion, places where science has never intruded. Not, however, for a lack of will. Richard Dawkins and Richard Feynman, for example, have little time for reverence. For them, the quest to understand predominates. To quote again from Feynman:

> *Today we cannot see whether Schrödinger's equation contains frogs, musical composers or morality, or whether it does not. We cannot say whether something beyond it like God is needed or not, and so we can all hold strong opinions either way.*

As science crosses the frontier where we start to understand complexity, the domain over which we may hold opinions may seem to diminish. Ahead of us lie demanding ethical challenges. The science journey of this century will require not only the courage of Odysseus, but every bit as much of his wisdom. And therein lies our greatest fear: will we have a future at all?

The world is not particularly worse than it was when Edmond Halley urged humanity to plan for the next transit of Venus. The world then was brutal indeed. As for people caring more for nature, this is doubtful. Indeed, as Halley wrote of 'the noblest sight' an English sailor was killing the last dodo on Mauritius. Now, against our best intentions, we are killing off species faster than ever. Robert May, until recently president of Britain's Royal Society, has estimated that, in antiquity, background extinctions corresponded to about one species every four years. Today we are running at 20,000 times that level.

And if we do not destroy the biosphere or bring about a catastrophe of global warming, what will we humans do to *ourselves*? As instant global media and internet reporting of localised tragedy beam into our homes and on to our computers, exposing our children more and more to the world's sadness, do we sow the seeds of despair? Open the door to a vast epidemic of depressive illness?

Throughout history, humanity has faced moral crises. This is the stuff of literature, from Homer to Shakespeare to Dostoevsky. But the dilemmas have grown more complex and dangerous. And suddenly we have a new fear, that the advance of western science and liberal thought that grew from the Renaissance may be swamped by a reaction from the dark ages, in which fanatical terrorism uses the opportunities presented by global connectedness to wreak havoc.

In short, we wonder: will the human race survive? Martin Rees, the current Astronomer Royal, has used the Copernican principle to look at our prospects. It goes like this:

*There is no special place in the universe, nor any special time.*

Ten percent of the people who have ever lived are alive today. Sixty billion have preceded us. So at what stage on the survival path of humanity do we stand? Consider a container with tickets numbered from 1 to 10, and another with tickets numbered from 1 to 1000. You choose a ticket from either container, not knowing which is which. You get the number 6. Now you can figure which container you dipped into. You are 100 times more likely to have dipped into the 1–10 scenario, than the 1–1000 scenario.

We are thus unlikely to be only a short way through the entire allotted span of the human race. Maybe we are halfway? Then, how long before we get through another 60 billion people? Probably not more than another century or two. Martin Rees figures our chance of surviving this century at no better than 50:50. I think that is a fair summary.

The human race has no divine right to continue to survive. But I want to live in a world where we are optimistic enough to believe that we *will* survive, that life will continue, and that we will hold fast to our humanity in the process. Science gives us no moral insight and advances no ethical principles; there are truths other than scientific truths, and in matters of ethics or human values the scientist's opinion has no higher status. But I believe science presents an abiding and self-consistent way of looking at the world, something solid, real and truly universal. That, in itself, is of inestimable value.

Science lets us see the way nature is: we experience this again as we glimpse the workings of the same great celestial clock James Cook saw when the transit of Venus began at the predicted moment on that June day in 1769. Science is the compass on the voyage we must all make into the twenty-first century. That we have it in our possession is astonishing. That we should never have found it, or that we should turn our back on it, is quite simply unthinkable.

***Paul Callaghan,** born in New Zealand, obtained a DPhil degree from Oxford University, working in low temperature nuclear physics. On his return to New Zealand he began researching the applications of magnetic resonance to the study of soft matter at Massey University. In 2001 he was appointed Alan MacDiarmid Professor of Physical Sciences at Victoria University of Wellington. He also heads the multi-university MacDiarmid Institute for Advanced Materials and Nanotechnology. Paul Callaghan has published around 220 articles in scientific journals, and a book on magnetic resonance. In 2001 he became a Fellow of the Royal Society of London.*

# Further Reading


**Search for the Lost Continent**

*Awesome Forces: The Natural Hazards that Threaten New Zealand,* Geoff Hicks and Hamish Campbell (eds): Te Papa Press, 1998

*The Chatham Islands: Heritage and Conservation,* Ian Atkinson et al: Canterbury University Press and the Department of Conservation, 1996

*The Earth: An Intimate History,* Richard Fortey: HarperCollins, 2004

*The Rise and Fall of the Southern Alps,* Glen Coates and Geoffrey Cox: Canterbury University Press, 2002

**The Road to Stonehenge**

*Hengeworld,* Mike Pitts: Arrow Books, 2001

*How to Gaze at the Southern Stars,* Richard Hall: Awa Press, 2004

*Matariki,* Libby Hakaraia: Reed Publishing, 2004

*Stonehenge Aotearoa,* Richard Hall, Kay Leather and Geoffrey Dobson: Awa Press, 2005

*Stonehenge Complete,* Christopher Chippendale: Thames & Hudson, 2004

*Work of the Gods,* Kay Leather and Richard Hall: Viking Seven Seas, 2004

**A Brilliant Civilisation**

*Ancient Voyagers in Polynesia,* Andrew Sharp: Angus & Robertson, 1963

*Kon-Tiki: Across the Pacific by Raft,* Thor Heyerdahl: Pocket, 1990

*The Kon-Tiki Expedition,* Thor Heyerdahl: Flamingo, 1996

*Vikings of the Sunrise,* Peter Buck: Greenwood Press, 1985

**TO THE FARTHEST ENDS OF THE EARTH**

*Discoveries: The Voyages of Captain Cook,* Nicholas Thomas: Allen Lane, 2003

*Eclipse,* Duncan Steel: Joseph Henry Press, 2001

*Endeavour: The Story of Captain Cook's First Great Epic Voyage,* Peter Aughton: Weidenfeld & Nicolson, 2002

*The Life of Captain James Cook,* J.C. Beaglehole: Stanford University Press, 1992

*The Transit of Venus,* Peter Aughton: Weidenfeld & Nicolson, 2004

**VOYAGING WITH COOK**

*Ancient Tahitian Society, Vols. I–III,* Douglas L. Oliver: Australian National University Press, 1974

*Science and Exploration in the Pacific: European Voyages to the Southern Oceans in the Eighteenth Century,* Margarette Lincoln, editor: The Boydell Press, 2001

*The Journals of Captain James Cook on His Voyages of Discovery: Vol. 1. The Voyage of the Endeavour 1768–1771,* J.C. Beaglehole: Cambridge University Press, 1955

*The Trial of the Cannibal Dog: Captain Cook in the South Seas,* Anne Salmond: Penguin, 2003

**TRAVELS IN TIME AND SPACE**

*How Nature Works,* Per Bak: Oxford University Press, 1997

*Infinite in All Directions,* Freeman Dyson: Harper & Row, 1988

*'In Praise of Science',* Lewis Wolpert: in *Science Today,* Routledge, 1997

*Our Final Century,* Martin Rees: Heinemann, 2003

*The Meaning of It All,* Richard Feynman: Persus Books, 1998

*The New Quantum Universe,* Tony Hey and Patrick Walters: Cambridge University Press, 2003

*The Scientists: A History of Science Told Through the Lives of Its Greatest Inventors,* John Gribbin and Adam Hook: Random House, 2004

*The Selfish Gene,* Richard Dawkins: Oxford University Press, 1976

*The Singularity is Near: When Humans Transcend Biology,* Ray Kurzweil: Viking, 2005

# Index

# More books from Awa Science

**The Elegant Universe of Albert Einstein**
*Tom Barnes et al. With an introduction by Rebecca Priestley*

---

In 1905 Albert Einstein is 26 and working at the Swiss Patent Office as a poorly paid technical expert. In his spare time he pursues his passion for physics. The papers he produces in this one year will change the world forever, explaining experimental findings that have puzzled scientists for decades, revolutionising the way we view the world, and laying the foundations of modern physics. In this book, leading scientists and historians explore the lead-up to Einstein's astonishing discoveries, and their world-shaking aftermath. *$24.99*

**North Pole, South Pole: The quest to understand Earth's magnetism**
*Gillian Turner*

---

In Roman legend a shepherd named Magnes, climbing a rocky hillside, was shocked to find that he could hardly move: his iron-studded boots and iron-tipped rod kept sticking to the rocks. Thus was recorded a mysterious natural phenomenon, magnetism, that would intrigue scientists and ordinary mortals. What was this invisible force? What caused it? Astonishingly, the answers would not be fully revealed until the advent of supercomputers. This absorbing book will make you marvel at the inner workings of our planet and its magnetic shield, without which life as we know it would be impossible. *$39.99*

**Zodiac: Remarkable stars on the path of the sun**
*Richard Hall*

---

At Stonehenge Aotearoa, where Richard Hall is resident astronomer, the most common question is, 'Tell me about my constellation' – referring to the inquirer's astrological star sign. This book is Hall's answer, a brilliant exposition of the history, science and mythology associated with the constellations that lie along the sun's path. Constellations are, in effect, no more than human constructs – patterns of bright stars as seen from Earth. But the zodiacal constellations have played a seminal role in the development of civilisations and religions, and in the eyes of some people even determine their future. *$34.99*

**The Awa Book of New Zealand Science**
*Rebecca Priestley, editor*

---

A landmark anthology of writings about New Zealand science and by New Zealand scientists that traces the development of all branches of science, from early naturalists' ramblings on flora and fauna, to geological accounts of Lake Rotomahana's famous pink and white terraces, to Ernest Rutherford splitting the atom, and modern breakthroughs in nanotechnology. This book is designed for the general reader and the pieces are highly accessible. Each has a brief introduction setting it in its historic and scientific context. *$44.99*

**How to Gaze at the Southern Stars**
*Richard Hall*

---

There is no more beautiful or intriguing sight than the night sky. From the beginning of human civilisation people have wondered about the twinkling lights, and the strange objects that occasionally fall to Earth. Come along with astronomer extraordinaire Richard Hall on a tour of the heavens as seen from the Southern Hemisphere. Hall weaves state-of-the-art science with the stories and myths of peoples across the globe and through the centuries. From the iconic Southern Cross to the mysterious dark companions of the Dog Stars, the night sky will never look the same again. *$24.99*

**Stonehenge Aotearoa: The Complete Guide**
*Richard Hall, Kay Leather, Geoff Dobson*

---

New Zealand outdoor observatory Stonehenge Aotearoa is unique nationally and internationally as a place of science and wonder. This full-colour guide takes you on a tour of the astonishing structure, and the prehistoric monuments which inspired its creation. Read about giant stones; constellations, eclipses and other wonders of the night sky; ancient and Maori astronomy, and Pacific navigation. Also included is fascinating information about the original Stonehenge in England, and many other prehistoric stone circles, statues and monuments around the world. *$19.99*